THE ESRI PRESS

*D*ictionary of
GIS
Terminology

THE ESRI PRESS

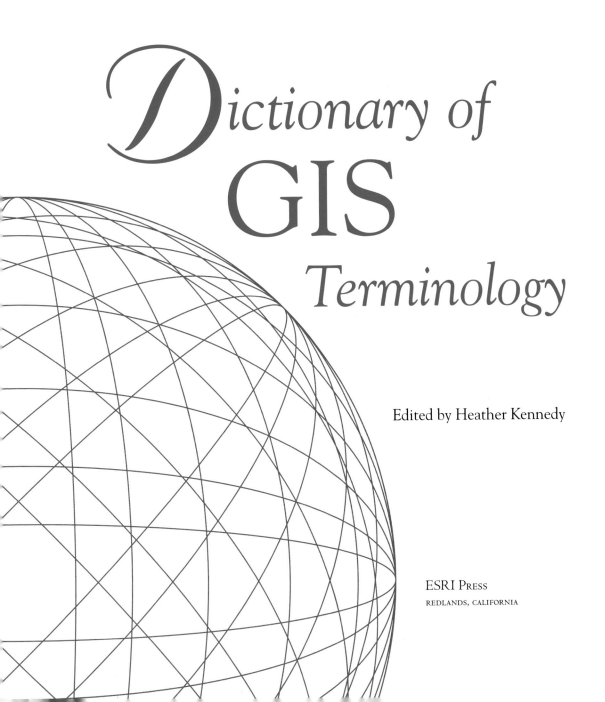

*D*ictionary of GIS *Terminology*

Edited by Heather Kennedy

ESRI PRESS
REDLANDS, CALIFORNIA

Environmental Systems Research Institute, Inc.
  The ESRI Press Dictionary of GIS Terminology
  ISBN 1-879102-78-1

First printing January 2001

Printed in the United States of America.

Library of Congress Cataloging-in-Publication Data
The ESRI Press dictionary of GIS terminology / edited by Heather Kennedy.
     p.   cm.
  ISBN 1-879102-78-1
  1. Geographic information systems—Dictionaries. I. Kennedy, Heather, 1965–
  G70.212 .E87 2001
  910'.285—dc21   00-013243

Published by Environmental Systems Research Institute, Inc., 380 New York Street, Redlands, California 92373-8100.

Books from ESRI Press are available to resellers worldwide through Independent Publishers Group (IPG). For information on volume discounts, or to place an order, call IPG at 1-800-888-4741 in the United States, or at 312-337-0747 outside the United States.

Contributors

Editors   Michael Karman, Gary Amdahl

Reviewers and consultants   Jonathan Bailey, David Barnes, Bob Booth, Hal Bowman, Patrick Brennan, Melissa K. Brenneman, Patricia Breslin, Clayton Crawford, Thomas A. Dunn, Cory L. Eicher, Matt Funk, Chuck Gaffney, Shelly Gill, R. W Greene, Erik Hoel, William E. Huxhold, Melita Kennedy, Michael Kennedy, Jonathan Makin, Norman T. Olsen, Tim Ormsby, Jaynya W. Richards, Mike Ridland, James TenBrink, Jennifer Wrightsell, Michael Zeiler, Mark D. Zollinger, Aaron Zureick

Book design, production, copyediting   Michael Hyatt

Cover design   Amaree Israngkura

Illustration   Steve Frizzell

AAT *See* arc attribute table.

abscissa [MATHEMATICS, COORDINATE GEOMETRY] In a rectangular coordinate system, the horizontal distance of the x-coordinate from the vertical or y-axis. For example, a point with the coordinates (7,3) has an abscissa of 7. The y-coordinate of a point is called the ordinate.

absolute accuracy [MAPPING] How well the position of an object on a map conforms to its location on the earth according to an accepted coordinate system such as geographic coordinates (latitude and longitude) or a State Plane coordinate system. *Compare* relative accuracy.

absolute coordinates [MAPPING, GPS] Coordinates that are referenced to the origin of a given coordinate system. *Compare* relative coordinates.

absolute location Also **absolute position** [MAPPING, GPS] The location of a point in geographic space with respect to an accepted coordinate system such as latitude and longitude.

access rights [COMPUTING] The privileges given to a user for reading, writing, deleting, and updating files on a disk or tables in a database. Access rights are stated as "no access," "read only," and "read/write."

accuracy The degree to which a value conforms to a specified standard for that value, or the degree to which a measured value is correct. *Compare* precision.

across-track scanner *See* whisk broom scanner.

active remote sensing Remote sensing systems, such as radar, that produce electromagnetic radiation and measure its reflection back from a surface. *Compare* passive remote sensing.

acutance [PHOTOGRAMMETRY, REMOTE SENSING] A measure, using a microdensitometer or other instrument, of how well a photographic system shows sharp edges between contiguous bright and dark areas.

A

address 1. Also **geocode** A point stored as an x,y location in a geographic data layer, referenced with a unique identifier. 2. [COMPUTING] A number that identifies a location in memory where data is stored. 3. A name identifying a site on the Internet or other network.

address geocoding Assigning x,y coordinates to tabular data such as street addresses or ZIP Codes so that they can be displayed as points on a map.

address matching Comparing addresses that identify the same location but which are recorded in different lists; used often as a precursor to address geocoding.

address range [GEOCODING] Street numbers running from lowest to highest along a street or street segment. Address ranges are generally stored as fields in the attribute table of a street data layer and are used for geocoding.

adjacency 1. The state or quality of lying close or contiguous. 2. [TOPOLOGY] The sharing of a side or boundary by two or more polygons.

adjacency analysis Also **contiguity analysis** [TOPOLOGY] Identifying and selecting geographic features that lie near or next to each other.

aerial photograph [REMOTE SENSING, PHOTOGRAMMETRY] A photograph of the earth's surface taken with a camera mounted in an airplane or balloon. Used in cartography to provide geographical information for basemaps.

aerial stereopair *See* stereopair.

affine transformation [GEOREFERENCING] A transformation that scales, rotates, and translates image or digitizer coordinates to map coordinates. In an affine transformation, the midpoint of a line segment remains the midpoint, all points on a line remain on that line, and parallel lines remain parallel.

air station Also **exposure station** [REMOTE SENSING, PHOTOGRAMMETRY] The location of the camera lens at the moment of exposure.

albedo [REMOTE SENSING] The ratio of the amount of electromagnetic energy reflected by a surface to the amount of energy striking it.

algorithm [MATHEMATICS] Any set of rules that can be followed to solve a complex problem, such as an encoded set of computer commands or the assembly instructions that come with a free-standing outdoor basketball goal.

alias In database management systems and on computer networks, an alternative name for someone or something. For example, a single e-mail alias may refer to a group of e-mail addresses.

aliasing [GRAPHICS] The jagged appearance of curves and diagonal lines on a raster display. Aliasing occurs when the detail of the diagonal line or curve exceeds the resolution of the pixels on the screen.

alidade [SURVEYING] 1. A telescope or peepsight mounted on a straightedge, used to measure direction. 2. The part of a theodolite containing the telescope and attachments.

allocation [GRAPH THEORY, NETWORK ANALYSIS] Assigning arcs or nodes in a network to the closest facility, until the capacity of the facility or each arc's limit of impedance is reached. For example, streets may be assigned to the nearest fire station, but only within a six-minute radius, or students may be assigned to the nearest school until it is full.

almanac 1. [GPS] File transmitted from satellites to receivers that contains information about the satellites' orbits. The receivers use the almanac to decide which satellite to track. 2. [ASTRONOMY, METEOROLOGY] An annual publication containing information on astronomical events and the daily movements of celestial bodies, used for navigation.

along-track scanner *See* push broom scanner.

altitude [SURVEYING, GEODESY] 1. The elevation above a reference datum, usually sea level, of any point on the earth's surface or in the atmosphere. 2. The z-value in a three-dimensional coordinate system.

AM/FM (Automated Mapping / Facilities Management) Automated cartography or geographic information systems (GIS) used by utilities and public works organizations for storing, manipulating, and mapping facility information such as pipe and road networks.

anaglyph [PHOTOGRAMMETRY] A composite picture made by superimposing two images of the same area. The images are displayed in complementary colors, usually red and green, and when viewed through filters of corresponding colors create a three-dimensional image.

analog Also **analogue** 1. An entity or variable represented continuously rather than in discrete steps; something that has value at any degree of precision. 2. [ELECTRONICS, COMPUTING] A continuously variable signal, or a circuit or device that carries such signals. *See also* discrete, digital.

analog display [GRAPHICS] A video display that presents an uninterrupted range of colors or gray shades. *Compare* digital display.

3

analog image [GRAPHICS] An image represented by continuous variation in tone, as in a photograph.

angular minute *See* minute.

angular units [SURVEYING, GEODESY] The unit of measurement on a sphere or a spheroid, usually degrees.

annotation [CARTOGRAPHY] 1. Text or graphics used to label the features in a geographic data layer. 2. Any explanatory text accompanying an image or map.

annotation

ANSI (American National Standards Institute) The United States government body responsible for approving U.S. industry standards in areas such as computing and communications. An ANSI standard is intended as a guide for manufacturers and consumers.

antipode [GEODESY, ASTRONOMY] That point on the surface of a globe or the earth which lies 180 degrees from a given point on the same surface.

anywhere fix [GPS] A position that a Global Positioning System (GPS) receiver can calculate without knowing its own location or the local time.

aphylactic projection A projection having neither equal area nor conformal characteristics. The term is rarely used.

apogee [ASTRONOMY, GPS] The point in the elliptical orbit of a terrestrial satellite that is farthest from the earth.

A

arc 1. An ordered string of x,y coordinate pairs (vertices) that begin at one location and end at another. Connecting the vertices creates a line. 2. A coverage feature class that represents linear features and polygon boundaries. One line feature can contain many arcs. Arcs are topologically linked to nodes (*see* arc–node topology) and to polygons (*see* polygon–arc topology). Their attributes are stored in an arc attribute table (AAT). *See also* node.

arc

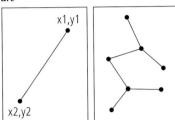

arc attribute table (AAT) A table containing attributes for arc coverage features. In addition to user-defined attributes, the AAT contains each arc's unique identifier, its from- and to-nodes, its left and right polygons, its length, and an internal sequence number. *See also* feature attribute table.

ARC GRID™ 1. An ArcInfo raster format for storing and displaying surface models. A grid partitions geographic space into square cells, each of which stores a numeric value. Values from sample data points are interpolated to create a continuous surface. 2. A program for assembling and disseminating global data sets for the United Nations and other agencies.

arc–node topology The data structure in a coverage used to represent linear features and polygon boundaries, and to support analysis functions such as network tracing. Nodes represent the beginning and ending vertices of each arc. Arcs that share a node are connected, and polygons are defined by a series of connected arcs. An arc that intersects another arc is split into two arcs. Each arc that defines all or part of a polygon boundary records the number of the polygon to its left and to its right, giving it a direction of travel. *See also* topology.

arc second [GEODESY] 1/3,600th of a degree (1 second) of latitude or longitude.

area 1. Also **polygon** A closed, two-dimensional shape defined by its boundary. 2. The size of a geographic feature measured in square units.

area chart A chart that emphasizes the difference between two or more groups of data; for example, the changes in a population from one year to the next. The area of interest is usually shaded in a solid color.

areal scale The ratio between the area of a feature on a map and the area of the same feature on the earth's surface. *See* scale.

argument 1. [COMPUTING] A value or expression passed to a function, command, or program. 2. [MATHEMATICS] An independent variable of a function.

ascending node [REMOTE SENSING] The point at which a satellite travelling south to north crosses the equator.

aspect The compass direction that a slope faces, usually measured clockwise from north.

aspect ratio The ratio of the width of an image to its height. A standard computer monitor aspect ratio is 4:3 (rectangular).

astrolabe [ASTRONOMY, NAVIGATION] An instrument that measures the vertical angle between a celestial body and the horizontal plane at an observer's position. The astrolabe was replaced by the sextant in the 15th century for marine navigation, but modern versions are still used to determine local time and latitude.

atlas [CARTOGRAPHY] A collection of maps organized around a theme, such as a world atlas, a national atlas, or a historical atlas.

atmospheric window [REMOTE SENSING] Regions of the electromagnetic spectrum in which radiation can be transmitted with relatively little interference from the atmosphere.

attenuation [REMOTE SENSING, PHOTOGRAMMETRY] The effects that atmospheric absorption and scattering have on light or other radiation that passes through the earth's atmosphere. Attenuation causes dimming and blurring in remotely sensed images.

attribute 1. Information about a geographic feature in a GIS, generally stored in a table and linked to the feature by a unique identifier. Attributes of a river might include its name, length, and average depth. *See* attribute table. 2. Cartographic information that specifies how features are displayed and labeled on a map; the cartographic attributes of the river in (1) above might include line thickness, line length, color, and font.

attribute table A table containing descriptive attributes for a set of geographic features, usually arranged so that each row represents a feature and each column represents one attribute. Each cell in a column stores the value of that column's attribute for that row's feature.

attribute table

Shape	Name	Population
Point	Dansville	114,234
Point	Portslain	77,265
Point	Bermisla	51,089
Point	Gold Ridge	39,172
Point	Shlener	30,422
Point	Cooper	19,963

authalic projection *See* equal-area projection.

automated cartography Cartography that uses plotters, software, and graphic displays to speed tasks traditionally associated with manual drafting. It does not involve spatial information processing. *Compare* geographic information system.

automation scale The scale at which nondigital data is made digital; for example, a map digitized at a scale of 1:24,000 has an automation scale of 1:24,000. The data can be rendered at different display scales.

AVHRR (Advanced Very High Resolution Radiometer) [REMOTE SENSING] A scanner flown on National Oceanic and Atmospheric Administration (NOAA) polar-orbiting satellites for measuring visible and infrared radiation reflected from vegetation, cloud cover, lakes, shorelines, snow, and ice. Used for weather prediction and vegetation mapping.

axis pl. **axes** 1. A line along which measurements are made in order to determine the coordinates of a location. 2. The line about which a rotating body turns. 3. In a spherical coordinate system, the line that directions are related to and from which angles are measured.

azimuth [GEOMETRY, NAVIGATION] The angle measured in degrees between a baseline drawn from a center point and another line drawn from the same point. Normally, the baseline points north and the angle is measured clockwise from the baseline.

azimuthal projection Also **true-direction projection, zenithal projection** A projection that preserves direction from its center, made by projecting the earth onto a tangent or secant plane. *See also* planar projection.

background image A satellite image or aerial photograph over which vector data is displayed. Although the image can be used to align coordinates, it is not linked to attribute information and is not part of the spatial analysis in a GIS.

backscatter [REMOTE SENSING] Electromagnetic energy that is reflected back toward its source by terrain or particles in the atmosphere.

backup [COMPUTING] A copy of one or more files made for safekeeping in case the originals are lost or damaged.

band 1. A set of adjacent wavelengths or frequencies with a common characteristic, such as the visible band in the electromagnetic spectrum. 2. One layer of a multispectral image that represents data values for a specific range of reflected light or heat, such as ultraviolet, blue, green, red, infrared, or radar, or other values derived by manipulating the original image bands. A standard color display of a multispectral image shows three bands, one each for red, green, and blue.

band pass filter [IMAGE PROCESSING, ELECTRONICS] A wave filter that allows signals in a certain frequency to pass through, while blocking or attenuating signals at other frequencies.

band separate An image format that stores each band of data in a separate file.

bandwidth 1. [PHYSICS, ELECTRONICS] A range within a band of wavelengths, frequencies, or energies, especially the range of frequencies required to transmit information at a specific rate. 2. [COMPUTING] Also **throughput** The amount of data that can flow through a communications channel, usually expressed in hertz for analog circuits and in bits per second (bps) for digital circuits. 3. [REMOTE SENSING] Also **spectral resolution** The range of frequencies that a satellite imaging system can detect.

bar/column chart A chart in which data values are represented by horizontal bars or vertical columns. The relative lengths of the bars or columns show differences and trends. Particularly effective where each individual data value has to be named.

barrier 1. A location in a linear network through which nothing can flow. 2. A line feature used to keep certain points from being used in the calculation of new values when interpolating a grid or creating a triangulated irregular network (TIN). A line can represent a cliff, a road, or any other interruption in the landscape. *See also* breakline.

bar scale Also **scale bar, graphic scale, linear scale** A line used to measure distance on a map, marked like a ruler in units proportional to the map's scale.

base data Map data over which other information is placed.

base-height ratio In aerial photography, the distance on the ground between the centers of overlapping photos, divided by aircraft height. In a stereomodel, base-height ratio is used to determine vertical exaggeration.

base layer A data layer in a GIS that all other layers are referenced to geometrically.

baseline 1. An accurately surveyed line used to measure other lines or the angles between them. 2. In a land survey system, a line passing east and west through the origin, used to establish township, section, and quarter-section corners. 3. [GPS] A pair of base stations that collect data simultaneously.

basemap A map depicting geographic features such as landforms, drainage, roads, landmarks, and political boundaries, used for locational reference and often including a geodetic control network as part of its structure. Examples include topographic and planimetric maps.

base station Also **reference station** A GPS receiver at a known location that broadcasts and collects correction information for roving GPS receivers. *See* differential correction.

bathymetry 1. The science of measuring and charting the depths of water bodies. 2. The measurements so obtained.

Bayesian estimate A spatial analysis technique that shows how the observed count of a variable differs from the mean or from an expected value.

bearing [SURVEYING, NAVIGATION] Also **bearing angle** The direction of a fixed point from a point of observation on the earth, expressed as an angle from a known direction, usually north, and usually measured from 0 degrees at the reference direction clockwise through 360 degrees. The terms bearing and azimuth are sometimes interchanged, but in navigation the former usually applies to objects on the earth while the latter applies to the direction of a point on the celestial sphere from a point on the earth.

bell curve *See* normal distribution.

benchmark A brass or bronze disk, set in a concrete base or similarly permanent structure, inscribed with a mark showing its elevation above or below an adopted datum.

Beziér curve [GRAPHICS] A curved line whose shape is derived mathematically. In graphics programs a Beziér curve usually has two endpoints and two control points that can be moved to change the direction and the steepness of the curve.

Beziér curve

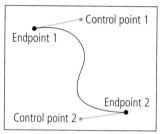

bilinear interpolation [DIGITAL IMAGE PROCESSING, ARC GRID] A technique used to resample raster data in which the value of each cell is calculated using the values of the four nearest cells. *See also* nearest neighbor assignment, cubic convolution.

binary 1. [MATHEMATICS] Base two. 2. [COMPUTING] Data with only two states, on/off, 0/1, true/false, or yes/no. 3. [COMPUTER USAGE] Digital data encoded as a sequence of bits but not as text; often used to describe machine readable code. 4. [PROGRAMMING] An operator that takes two arguments.

biogeography The study of the geographical distribution of animals and plants.

bit image *See* bit map.

bit map Also **bit image** An image format in which each pixel on the screen is represented by one or more bits. The number of bits per pixel determines the shades of gray or number of colors that a bit map can represent.

BLOB (Binary Large Object) 1. A large block of data such as an image, a sound file, or geometry, stored in a database. The database cannot read the BLOB's structure and only references it by its size and location. 2. The data type of the column in the database that stores said BLOB.

block *See* census block.

block group [DEMOGRAPHY] A geographical area that combines adjacent census blocks into a group of approximately one thousand people.

Boolean expression Also **logical expression** An expression that reduces to a true or false condition, for example, HEIGHT < 70 AND DIAMETER = 100. *See also* Boolean operator.

Boolean operator Also **logical operator** A word that combines simple logical expressions into a complex expression. The four most common in programming use are AND (logical conjunction), OR (logical inclusion), XOR (exclusive or), and NOT (logical negation).

border arcs 1. The arcs that create the boundary of a polygon coverage. 2. In ARC/INFO LIBRARIAN™, the arcs that split a polygon coverage into tiles.

boundary line A line between politically defined territories, such as states or countries. Boundaries between privately owned land parcels are usually called property lines.

boundary monument An object that marks an accurately surveyed position on or near a boundary line.

boundary survey 1. Also **boundary plat** A map that shows property lines and corner monuments of a parcel of land. 2. The survey taken to gather the data for such a map.

bounding rectangle The rectangle defined by one or more geographical features in coordinate space, determined by the minimum and maximum coordinates in the x and y directions.

breaklines Linear features in a TIN that are enforced as triangle edges. Hard breaklines represent distinct interruptions in the slope of a surface, such as roads or streams. Soft breaklines are generally used to maintain known z-values or the edges of line and polygon features.

brightness theme A grid theme whose cell values are used to vary the brightness of another grid theme. Most commonly, hillshade grids are used as brightness themes for elevation grids. The effect is to display the elevation surface in relief.

B

buffer 1. [TOPOLOGY] A polygon enclosing a point, line, or polygon at a specified distance. 2. [COMPUTING] A storage area, usually in RAM, that holds data while it is transferred from one location to another.

buffers

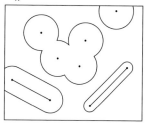

Build An ArcInfo™ command that constructs topology and creates a feature attribute table for a coverage. Unlike Clean, it does not create intersections or correct undershoots and overshoots. *Compare* Clean.

C

C/A (Coarse/Acquisition) code Also **Civilian code, S-code** [GPS] The standard pseudo-random code used by most civilian GPS receivers. *Compare* P-code.

CAD *See* computer-aided design.

cadastral survey A boundary survey taken for the purposes of taxation.

cadastre A public record of the dimensions and value of land parcels, used to record ownership and calculate taxes.

calibration 1. Comparing the accuracy of an instrument's measurements to a known standard. 2. In spatial analysis, choosing attribute values and computational parameters so that a model properly represents the situation being analyzed. For example, in pathfinding and allocation, calibration generally refers to assigning or calculating impedance values.

candidate key In a relational database, any key that can be used as the primary key in a table. *See also* primary key.

cardinal direction [NAVIGATION] One of the astronomical directions on the earth's surface: north, south, east, and west.

cardinality 1. The number of elements in a mathematical set. 2. In a relationship between objects in a database, the number of objects of one type that are associated with objects of another type. A relationship can have a cardinality of one-to-one, one-to-many, or many-to-many.

cardinal points [NAVIGATION] The four cardinal directions, indicated on a compass.

carrier [PHYSICS, GPS] An electromagnetic wave, such as radio, whose modulations are used as signals to transmit information.

carrier-aided tracking [GPS] Signal processing that uses the GPS carrier signal to lock onto the pseudo-random code generated by the satellite.

carrier phase GPS GPS measurements that are calculated using the carrier signal of the satellite. *Compare* code phase GPS.

Cartesian coordinate system [GEOMETRY] A system of reference in which location is measured along the planes created by two or three mutually perpendicular intersecting axes. In two dimensions, points are described by their positions in relation to two axes, x and y. A third axis, z, is added to measure locations in three dimensions. Relative measure of distance, area, and direction are constant throughout the system. Named after René Descartes, who originated the two-dimensional system in the seventeenth century.

Cartesian coordinate system

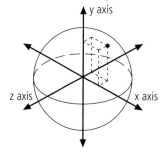

cartogram A diagram or abstract map in which geographical areas are exaggerated or distorted in proportion to the value of an attribute.

cartographic elements The primitive components that make up a map, such as the neatline, legend, scale, titles, and figures.

cartographic license The extent to which a cartographer can change the appearance, layout, and content of a map without making it less accurate.

cartography The design, compilation, drafting, and reproduction of maps.

cartouche An ornamental frame around a map, often including the title or legend. Rarely used on modern maps.

celestial sphere The sky considered as the inside of a sphere of infinitely large radius that surrounds the earth, on which all celestial bodies except the earth are imagined to be projected.

cell 1. The smallest square in a grid. Each cell usually has an attribute value associated with it. 2. A pixel.

cell size Also **pixel size** The area on the ground covered by a single pixel in an image, measured in map units.

census block [DEMOGRAPHY] The smallest geographic unit used by the U.S. Census Bureau for reporting census data and for generating geographic base files such as DIME and TIGER® files. A block is enclosed by any natural or human-made features that form a logical boundary, such as roads, political boundaries, or shorelines.

census tract A geographical area that combines adjacent census blocks into a group of approximately four thousand people.

center 1. The point on a circle or in a sphere equidistant from all other points on the object. 2. The point from which angles or distances are measured. 3. In network allocation, a location from which resources are distributed or to which they are brought.

centerline A line digitized along the center of a linear geographic feature, such as a street or a river, that at a large enough scale would be represented by a polygon.

central meridian [MAPPING, NAVIGATION] The line of longitude that defines the center and often the x origin of a projected coordinate system.

centroid 1. The geometric center of a figure. Of a line, it is the midpoint; of a polygon, the center of area; of a three-dimensional figure, the center of volume. 2. The center of mass of a line, polygon, or three-dimensional figure. For example, the population center of an area could be calculated as its center of mass, using population density as the weight.

chain 1. [SURVEYING] A unit of length equal to 66 feet, used especially in U.S. public land surveys. Ten square chains equal one acre. 2. *See* arc.

character A letter, a digit, or a special graphic symbol (e.g., *, |, -) treated as a single unit of data and usually stored as one byte.

chart 1. A map for air or water navigation. 2. Also **graph** A diagram showing the relation between two or more variable quantities, usually measured along two perpendicular axes.

charts

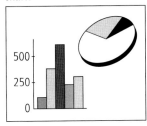

chord A straight line that joins two points on a curve.

C

choropleth A thematic map in which areas are colored or shaded to reflect the density of the mapped phenomenon or to symbolize classes within it.

choropleth

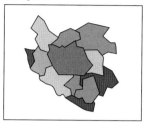

chroma The saturation, purity, or intensity of a color. *See* hue, value.

chronometer [NAVIGATION] An extremely accurate clock that remains accurate through all conditions of temperature and pressure. Developed in the 18th century by John Harrison; used at sea for determining longitude.

Clarke Belt An orbit 22,245 miles (35,800 kilometers) above the equator in which a satellite travels at the same speed that the earth rotates. Named after writer and scientist Arthur C. Clarke.

Clarke spheroid (ellipsoid) of 1866 A reference spheroid having a semimajor axis of approximately 6,378,206.4 meters and a flattening of 1/294.9786982. The Clarke spheroid is the basis for NAD 1927 and other datums.

class 1. A group or category of attribute values. 2. Pixels in a raster file that represent the same condition.

classification Grouping items into categories.

Clean An ArcInfo command that generates a coverage with correct polygon or arc–node topology by adjusting geometric coordinate errors, creating intersections, assembling arcs into polygons, and creating feature attribute information for each polygon (a polygon attribute table, or PAT) or arc (an arc attribute table, or AAT). *Compare* Build.

clean data Data that is free from error.

cleaning Also **scrubbing** Improving the appearance of scanned or digitized data by correcting overshoots and undershoots, making lines thinner or thicker, closing polygons, and so forth.

clearinghouse A place that stores and disseminates data.

client/server A software system with a central processor (server) that accepts requests from multiple users (clients).

clinometric map Also **slope map** A map that shows steepness with colors or shading.

Clip [GEOPROCESSING] An ArcInfo command that extracts the features from one coverage that reside entirely within a boundary defined by features in another coverage (called the clip coverage).

Clip

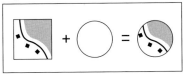

CMYK A color model that combines the printing inks cyan, magenta, yellow, and black to create a range of other colors. Most commercial printing uses this color model.

code phase GPS GPS measurements calculated using the pseudo-random code (C/A or P) transmitted by a GPS satellite. *Compare* carrier phase GPS.

color composite A color image made by assigning a different color to each of the separate monotone bands of a multispectral image and then superimposing them.

color model Any system that organizes colors according to their properties. Examples include RGB (red, green, blue), CMYK (cyan, magenta, yellow, black), HSB (hue, saturation, brightness), HSV (hue, saturation, value), HLS (hue, lightness, saturation), and CIE-L*a*b (Commission Internationale de l'Eclairage-luminance, a, b).

color ramp A range of colors used to show ranking or order among classes on a map.

color separation 1. Preparing a separate printing plate for each color used in producing a map or chart. 2. Scanning a map with color filters to separate the original image into single color negatives.

column Also **field, item** The vertical dimension of a table. Each column stores the values of one type of attribute for all of the records, or rows, in the table. All of the values in a given column are of the same data type; e.g., NUMBER, STRING, BLOB, DATE. *See* attribute table.

column chart *See* bar/column chart.

command An instruction to a computer program, usually one word or concatenated words or letters, issued by the user from a control device such as a keyboard or read from a file by a command interpreter. Menu items on a GUI are also often referred to as commands.

command-line interface [COMPUTING] An on-screen interface in which the user types in commands at a prompt. *Compare* GUI.

compass 1. [NAVIGATION] A round flat instrument that marks the cardinal directions around its edge and houses a floating magnetic needle that pivots to magnetic north. 2. An instrument with two legs connected by a joint, used to draw and measure circles. *See* cardinal direction, compass point, compass rose.

compass point One of the thirty-two divisions into which the circle around the needle of a compass is divided, each equal to 11.25 degrees.

compass rose Also **wind rose** A small compass drawn on a map or navigational chart, subdivided clockwise from 0 degrees to 360 degrees with 0 indicating true north. On older maps and charts it was a decorated diagram of cardinal directions, divided into sixteen or thirty-two points. Originally called *rosa ventorum*, or "rose of the winds."

compass rose

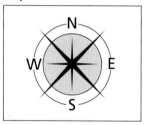

complex polygon A polygon that has inner as well as outer boundaries, that is, holes or islands.

computer-aided design (CAD) An automated system for the design, drafting, and display of graphical information. *Compare* geographic information system.

computer-aided drafting design (CADD) *See* computer-aided design.

computer-aided mapping (CAM) Mapping software that typically uses a vector format, but without topological links. *Compare* geographic information system.

concatenate To join two or more character strings together, end to end. For example, to combine the two strings "howdy" and "stranger" into the single string "howdy stranger."

concatenated key In a relational database table, a primary key made by combining two or more keys that together form a unique identifier.

conditional operator A symbol or keyword that specifies the relationship between two values. Conditional operators are used to query a database. Examples include = (equal to), < (less than), > (greater than).

conditional statement A programming language statement that executes one option if the statement is true, another if it is false. The If-Then-Else statement is an example of a conditional statement.

conflation A set of procedures that aligns the features of two geographic data layers and then transfers the attributes of one to the other. *See also* rubber sheeting.

conformality The characteristic of a map projection that preserves the shape of any small geographical area.

conformal projection Also **orthomorphic projection** A projection that preserves the correct shapes of small areas. Graticule lines intersect at 90-degree angles, and at any point on the map the scale is the same in all directions. A conformal projection maintains all angles, including those between the intersections of arcs; therefore the size of areas enclosed by many arcs may be greatly distorted. No map projection can preserve the shapes of larger regions.

conic projection A projection made by projecting geographic features onto a tangent or secant cone that is wrapped around the globe in the manner of a party hat. The cone is then cut and unrolled into a flat map.

conjoint boundary A boundary shared by two geographical areas or map sheets.

connectivity [TOPOLOGY] How geographic features in a network of lines are attached to one another functionally or spatially. *See also* arc–node topology.

connectivity analysis Identifying areas or points that are, or are not, connected to other areas or points by tracing routes along linear features.

connectivity rules [TOPOLOGY] Rules that constrain the type and number of network features that can be connected to one another in a geodatabase.

constant azimuth *See* rhumb line.

containment The relationship between a feature or a set of features and a polygon that completely surrounds them, in the same layer or different layers.

conterminous Also **coterminous** Having the same or coincident boundaries. *See also* contiguity.

contiguity 1. The state of lying next to or close to one another. 2. [TOPOLOGY] The identification of adjacent polygons by recording the left and right polygon for each arc in a geographic layer. *See also* polygon–arc topology.

continuous data Data such as surface elevation or temperature that varies without discrete steps. Since computers store data discretely, continuous data is usually represented by TINs, rasters, or contour lines, so that any location has either a specified value or one that can be derived. *See also* interpolation.

continuous tone image A photograph that has not been screened and so displays all the tones from black to white or dark to light color. *See also* halftone image, dot screen.

contour interval The difference in elevation between two contour lines.

contour line A line drawn on a map connecting points of equal elevation above a datum, usually mean sea level.

contour lines

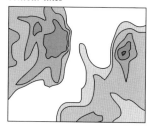

contour tagging Assigning values to scanned contour lines.

contrast [REMOTE SENSING, PHOTOGRAPHY, PHOTOGRAMMETRY] The ratio between the energy emitted or reflected by an object and that emitted or reflected by its immediate surroundings.

contrast ratio The ratio between the maximum and the minimum brightness values in an image.

contrast stretch [DIGITAL IMAGE PROCESSING] Increasing the contrast in an image by expanding its grayscale range to the range of the display device.

control *See* ground control.

convergence angle Also **meridional convergence** The angle between a vertical line (grid north) and true north on a map.

cookie-cut [GEOPROCESSING] A spatial operation that excludes the area outside a particular zone. *See* Clip.

Coordinated Universal Time (UTC) *See* Greenwich mean time.

coordinate geometry (COGO) Automated mapping software used in land surveying that calculates locations using distances and bearings from known reference points.

coordinates 1. The x- and y-values that define a location in a planar coordinate system. 2. The x-, y-, and z-values that define a location in a three-dimensional coordinate system.

coordinate system A reference system consisting of a set of points, lines, and/or surfaces, and a set of rules, used to define the positions of points in space in either two or three dimensions. *See also* geocentric coordinate system, geographic coordinate system, planar coordinate system.

coordinate transformation Also **rectification** Converting the coordinates in a map or an image from one system to another, typically through rotation and scaling.

corridor A buffer drawn around a linear feature.

corridor analysis Buffer analysis usually applied to environmental and land-use data in order to find the best locations for building roads, pipelines, and other linear transportation features.

coterminous *See* conterminous.

coverage An ArcInfo vector data storage format. A coverage stores the location, shape, and attributes of geographic features, and usually represents a single theme such as soils, streams, roads, or land use. Map features are stored as both primary features (e.g., arcs, polygons, and points) and secondary features (e.g., tics, links, and annotation). The attributes of geographic features are stored independently in feature attribute tables.

coverage

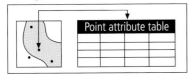

C

coverage units The units of the coordinate system in which a coverage is stored (e.g., feet, meters, inches).

credits Acknowledgement on a map of its author, its sources, and the date it was made.

cross tabulation Comparing attributes in different map layers according to location.

cross-tile indexing Indexing features that cross tile boundaries in a map library by storing them as one or more features in each tile, instead of storing them each as a single feature.

cubic convolution [DIGITAL IMAGE PROCESSING, ARC GRID] A technique used to resample raster data in which the value of each cell is calculated using the values of the sixteen nearest cells. *See also* bilinear interpolation, nearest neighbor assignment.

cultural features Human-made features, on a map or on the ground.

cultural geography Geography that studies human culture and its effects on the earth.

curve fitting Converting short connected straight lines into smooth curves to represent features such as rivers, shorelines, and contour lines. The curves that result pass through or close to the existing points.

cycle 1. [REMOTE SENSING] One oscillation of a wave. 2. [NETWORK ANALYSIS] A path or tour beginning and ending at the same location. 3. A set of lines forming a closed polygon.

cylindrical projection A projection made by projecting geographic features onto a tangent or secant cylinder wrapped around the globe. The cylinder is then cut and unrolled into a flat map.

dangle length Also **dangle tolerance** For an ArcInfo coverage, the minimum length allowed for dangling arcs during the Clean process. Clean removes dangling arcs shorter than the dangle length.

dangling arc An arc having the same polygon on both its left and right sides and having at least one node that does not connect to any other arc. It often identifies a location where arcs do not connect properly (an undershoot), or where an arc was digitized past its intersection with another arc (an overshoot). A dangling arc is not always an error; for example, it can represent a cul-de-sac in a street network. *See also* dangling node.

dangling arcs

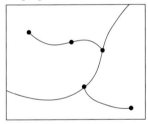

dangling node The endpoint of a dangling arc.

data Any collection of related facts arranged in a particular format; often, the basic elements of information that are produced, stored, or processed by a computer.

data automation Any electronic, electromechanical, or mechanical means for recording, communicating, or processing data.

database One or more structured sets of persistent data, managed and stored as a unit and generally associated with software to update and query the data. A simple database might be a single file with many records, each of which references the same set of fields. Examples of popular databases include Sybase®, dBASE®, Oracle®, and INFO™. A GIS database includes data about the spatial locations and shapes of geographic features recorded as points, lines, areas, pixels, grid cells, or TINs, as well as their attributes.

D

database lock A mechanism that prevents conflicting access to a database when several people are using it at once.

database management system (DBMS) A set of computer programs that organizes the information in a database and provides tools for data input, verification, and storage.

data capture Any operation that converts digital or analog data into computer-readable form. Geographic data can be downloaded directly into a GIS from sources such as remote sensing or GPS, or it can be digitized, scanned, or keyed in manually from paper maps or photographs.

data conversion Translating data from one format to another, usually in order to move it from one system to another.

data definition language (DDL) SQL statements that can be used either interactively or within a programming language to create a new database, set permissions on it, and define its attributes.

data dictionary [METADATA] A set of tables containing information about the data stored in a GIS database, such as the full names of attributes, meanings of codes, scale of source data, accuracy of locations, and map projections used.

data entry The transfer of data into a computer by manual key entry.

data file A file that holds text, graphics, or numbers. *Compare* executable file.

data format The structure used to store a file or record.

data integration Combining databases or data files from organizations that collect information about the same entities (such as properties, census tracts, or sewer lines). Doing so prevents redundant work and creates new ways to analyze the information.

data logger Also **data recorder** A lightweight, hand-held field computer used to store data collected by a GPS receiver.

data marker A column, bar, area, point symbol, or pie slice in a chart that represents tabular data.

data message [GPS] Information in a satellite's GPS signal that reports its orbital position, operating health, and clock corrections.

data set Any collection of data with a common theme.

data type 1. In a database table, the types of data that columns and variables can store. Examples include character, floating point, and integer. 2. [PROGRAMMING] Specifications of the possible range of values of a data set, the operations that can be performed on it, and the way the values are stored in memory.

datum [GEODESY, SURVEYING] In the most general sense, any set of numeric or geometric constants from which other quantities, such as coordinate systems, can be defined. There are many types of datums, but most fall into two categories: horizontal and vertical. *See* geodetic datum, geocentric datum, horizontal control datum, vertical control datum.

datum plane Also **datum level, reference level** A surface to which heights, elevations, or depths are referenced.

DBMS *See* database management system.

DDL *See* data definition language.

dead reckoning A navigation method of last resort that uses the most recently recorded position of a ship or aircraft, along with its speed and drift, to calculate a new position.

decimal degrees Degrees of latitude and longitude expressed in decimals instead of in degrees, minutes, and seconds. Decimal degrees are computed with the formula
$$decimal\ degrees = degrees + minutes/60 + seconds/3,600$$
Using this formula, 73° 59' 15" longitude is equal to 73.9875 decimal degrees.

declination 1. In a spherical coordinate system, the angle between the equatorial plane and a line to a point somewhere on the sphere. 2. The arc between the equator and a point on a great circle perpendicular to the equator. 3. [ASTRONOMY] The angular distance between a star or planet and the celestial equator. 4. Magnetic declination. The horizontal angle between geographic north and magnetic north from the point of observation.

degree A unit of angular measure, represented by the symbol °. The circumference of a circle contains 360 degrees.

degrees/minutes/seconds (DMS) A measurement of degrees of latitude and longitude in which each degree is divided into sixty minutes and each minute is divided into sixty seconds.

degrees/minutes/seconds

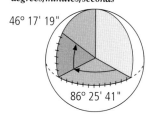

delimiter A character, such as a space or a comma, that separates words or values.

DEM *See* digital elevation model.

demographics The statistical characteristics (such as age, birth rate, and income) of a human population.

demography The study of human vital and social statistics, such as births, deaths, health, marriage, and welfare.

densify To add vertices to a line at specified distances, without altering the line's shape. *Compare* spline.

densitometer [PHOTOGRAMMETRY] An instrument for measuring the opacity of translucent materials such as photographic negatives and optical filters.

density slicing [DIGITAL IMAGE PROCESSING] A technique normally applied to a single-band monochrome image for highlighting areas that appear to be uniform in tone, but are not. Grayscale values (0–255) are converted into a series of intervals, or slices, and different colors are assigned to each slice. Often used to highlight variations in vegetation.

depression contour Also **hachured contour** [CARTOGRAPHY] A contour line indicating a closed depression on a topographic map, usually drawn with tick marks, or hachures, along the inside of the lower area.

depth curve Also **depth contour, bathymetric curve** A line on a map connecting points of equal depth below the hydrographic datum.

descending node [REMOTE SENSING] The point at which a satellite traveling north to south crosses the equator.

descriptor *See* attribute.

desktop GIS Mapping software that runs on a personal computer and can display, query, update, and analyze geographic locations and the information linked to those locations.

desktop mapping Mapping software for personal computers, ranging from systems that can only display data to full geographic information systems. *See* desktop GIS.

developable surface [CARTOGRAPHY] A geometric shape such as a cone, cylinder, or plane that can be flattened without being distorted. Many map projections are classified in terms of these shapes.

device coordinates The coordinates on a digitizer or a display, as opposed to those of a recognized datum or coordinate system.

DGPS Differential Global Positioning System. *See* differential correction.

diazo process [CARTOGRAPHY] A way of quickly and inexpensively copying maps using a diazo compound, ultraviolet light, and ammonia.

difference image [DIGITAL IMAGE PROCESSING] An image made by subtracting the pixel values in one image from those in another.

differential correction A technique for increasing the accuracy of GPS measurements by comparing the readings of two receivers, one roving, the other fixed at a known location.

Differential Global Positioning System *See* differential correction.

digital [COMPUTING] Also, often, **binary** Data processed in discreet, quantified units. Most computers process information as combinations of binary digits, or bits.

digital count The total number of pixels for each data value in an image.

digital display A video display that shows values as arrays of numbers. It can display only a finite number of colors. *Compare* analog display.

digital elevation model (DEM) Also **digital terrain model (DTM)** 1. The representation of continuous elevation values over a topographic surface by a regular array of z-values, referenced to a common datum. Typically used to represent terrain relief. 2. The database for elevation data by map sheet from the National Mapping Division of the U.S. Geological Survey.

digital image [REMOTE SENSING, PHOTOGRAMMETRY, GRAPHICS] An image stored in binary form and divided into a matrix of pixels, each of which consists of one or more bits of information that represent either the brightness, or the brightness and color, of the image at that point.

digital image processing (DIP) Any technique that changes the digital values of an image for the sake of analysis or enhanced display, such as density slicing or low- and high-pass filtering.

digital line graph (DLG) Vector data files of transportation, hydrography, contour, and public land survey boundaries from USGS basemaps.

digital number (DN) A value assigned to a pixel in a digital image.

digital orthophoto *See* orthophotograph.

digital terrain model (DTM) *See* digital elevation model.

digitize To convert the shapes of geographic features from media such as paper maps or raster imagery into vector x,y coordinates. *See* digitizer.

digitizer 1. (Manual) A device consisting of a tablet and a handheld cursor that converts electronic signals from positions on the tablet surface to digital x,y coordinates, yielding vector data consisting of points, lines, and polygons. 2. The title of the person who uses a digitizer. 3. (Video) An optical device that translates an analog image into an array of digital pixel values. A video digitizer can be used in place of a manual digitizer, but since it produces a raster image, additional software must be used to convert the data into vector format before topological analysis can be done.

digitizer

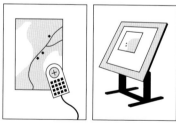

Dijkstra's algorithm An algorithm that uses the topological structure in a network to find the shortest path between two points.

Dilution of Precision (DOP) Also **Geometric Dilution of Precision (GDOP)** [GPS] An indicator of satellite geometry for a constellation of satellites used to determine a position. Positions with a higher DOP value generally constitute poorer measurement results than those with lower DOP. Factors determining the total GDOP for a set of satellites include, to name a few, PDOP (Positional DOP), HDOP (Horizontal DOP), VDOP (Vertical DOP), and TDOP (Time DOP).

DIME *See* GBF/DIME.

DIP *See* digital image processing.

directed link A line between two nodes with one direction specified.

directed network A network in which each line has an associated direction of flow.

directional filter [DIGITAL IMAGE PROCESSING] An edge-detection filter that enhances linear features in an image that are oriented in a particular direction.

directory An area of a computer disk that holds a set of data files and/or other directories. Directories are arranged in a tree structure, in which each branch is a subdirectory of its parent branch. The location of a directory is specified with a pathname, for example C:\gisprojects\shrinkinglemurhabitat\grids.

Dirichlet tessellations *See* Thiessen polygons.

discrete data Also **integer data** Geographic features that are represented by points, lines, or bounded polygons. *Compare* continuous data.

display resolution The number of pixels displayed on a monitor, measured horizontally and vertically (for example, 1,024 by 768).

display scale The scale at which digital data is rendered on a computer screen or on a printed map. *Compare* automation scale.

D

Dissolve 1. [GEOPROCESSING] An ArcInfo command that removes boundaries between adjacent polygons that have the same value for a specified attribute. 2. Removing unnecessary boundaries between features after data has been captured, such as the edges of adjacent map sheets.

Dissolve

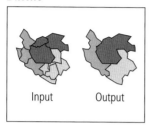

Input   Output

distance The amount of space between two things that may or may not be connected, such as two points. Differentiated from length, which always implies a physical connection.

distance decay 1. A mathematical representation of the effect of distance on the accessibility of locations and the number of interactions between them, reflecting the notion that demand drops as distance increases. It can be expressed either as a power function or as an exponential function. 2. The property by which two nearby points have more in common than two distant points.

distance units The units (feet, miles, meters, or kilometers) that ArcMap™ and ArcView® use to report measurements, dimensions of shapes, and distance tolerances and offsets. *Compare* map units.

distortion On a map or an image, the misrepresentation of shape, area, distance, or direction of or between geographic features when compared to their true measurements on the curved surface of the earth.

distribution 1. The amount or frequency of the occurrence of a thing or things within a given area. 2. The set of probabilities that a variable will have a particular value.

dithering 1. [GRAPHICS] Creating new shades by interspersing pixels of different colors. 2. [GPS] The introduction of digital noise to a GPS signal, used by the U.S. Department of Defense to make positions gathered by GPS receivers less accurate.

diurnal Daily, as in the revolution of the earth.

diurnal arc The apparent path from rise to set made by a heavenly body across the sky.

DLG *See* digital line graph.

DMS *See* degrees/minutes/seconds.

domain The range of values allowed for a column in a database.

DOP *See* Dilution of Precision.

Doppler-aided Also **Doppler-aiding** [GPS] Signal processing that uses a measured Doppler shift to help the receiver track the GPS signal.

Doppler shift Also **Doppler effect** The apparent change in frequency of sound or light waves caused by the relative motion between a source and an observer. As they approach one another, the frequency increases; as they draw apart the frequency decreases.

dot distribution map A map that uses dots or other symbols to represent the presence, quantity, or value of a thing in a specific area. Symbols whose sizes differ in relation to the phenomenon being mapped are called proportional symbols.

dot pattern A matrix of dots that approximates changing values of brightness in a printed image.

dot screen [CARTOGRAPHY, GRAPHICS] A photographic film covered with uniformly sized, evenly spaced dots used to break up a solid color, producing an apparently lighter color.

dots per inch (DPI) A measure of the resolution of scanners, printers, and graphic displays. The more dots per inch, the more sharply an image is represented. Desktop printers, for example, usually have resolutions ranging from 300 to 600 DPI, while commercial printing typically uses resolutions of 1,200 to 2,400 DPI.

double precision A high level of coordinate accuracy based on the number of significant digits that can be stored for each coordinate.

Douglas–Poiker algorithm Also **Douglas–Peucker algorithm** A formula that simplifies complex line features by reducing the number of points used to represent a digitized line.

downstream In network tracing, the direction along a line that is the same as the direction of flow. Direction of flow is determined by a user-defined convention. *See also* directed network.

DPI *See* dots per inch.

drafting Cartographic reproduction by way of pencil, pen and ink, or scribing.

D

drainage [CARTOGRAPHY] All features on a map associated with water, such as rivers, lakes, and shorelines.

drape A perspective or panoramic rendering of a two-dimensional image superimposed on a three-dimensional surface.

drum plotter *See* plotter.

drum scanner A rotating cylinder across which a sensor beam moves rapidly. Maps are scanned and converted to digital format as they rotate on the drum. *Compare* flatbed scanner.

DTM Digital terrain model. *See* digital elevation model.

Dual Independent Map Encoding *See* GBF/DIME.

dynamic segmentation Computing the locations of events along linear features, such as accidents on a highway or change in water quality along a river. Dynamic segmentation associates multiple sets of attributes with any portion of a linear feature at run time without affecting the underlying x,y coordinate data. *See also* route, measure, run time.

ℰ

easting 1. The distance east that a point in a coordinate system lies from the origin, measured in that system's units. 2. The x-value in a rectangular coordinate system.

eccentricity Also **ellipticity** [GEOMETRY] A measure of how much an ellipse deviates from a circle, expressed as the ratio of the distance between the center and one focus of an ellipsoid to the length of its semimajor axis. The square of the eccentricity, e^2, is commonly used with the semimajor axis a to define a spheroid in map projection equations.

ecliptic 1. The great circle formed by the intersection of the plane of the earth's orbit around the sun (or apparent orbit of the sun around the earth) and the celestial sphere. 2. The mean plane of the earth's orbit around the sun.

edge 1. In a TIN, a line that connects two nodes. 2. In a geodatabase geometric network, a line that connects two junctions. 3. In an image, the margin between areas of different tones or colors.

edge detection [DIGITAL IMAGE PROCESSING] A technique for isolating optical edges in a digital image by examining it for abrupt changes in pixel value.

edge enhancement [DIGITAL IMAGE PROCESSING] A technique for emphasizing the appearance of edges and lines in an image. *See also* high-pass filter.

edgematching [GEOPROCESSING] Assigning the correct coordinate and attribute information to geographic features that connect across the boundaries of adjacent data layers.

elastic transformation *See* rubber sheeting.

electromagnetic radiation Energy that moves through space at the speed of light as different wavelengths of time-varying electric and magnetic fields. Types of electromagnetic radiation include gamma, x, ultraviolet, visible, infrared, microwave, and radio.

electromagnetic spectrum The entire range of wavelengths or frequencies over which electromagnetic radiation extends.

electronic atlas A mapping system that displays but does not analyze geographic data.

33

elevation Also **altitude, height** The vertical distance of a point or object above or below a reference surface or datum (generally mean sea level).

ellipsoid 1. [GEOMETRY] A closed surface all planar sections of which are ellipses. An ellipsoid has three independent axes, and is usually specified by the lengths a,b,c of the three semi-axes. If an ellipsoid is made by rotating an ellipse about one of its axes, then two of the axes of the ellipsoid are the same, and it is called an ellipsoid of revolution or spheroid. If the lengths of all three of its axes are the same, it is a sphere. 2. Also **spheroid, oblate spheroid** [GEODESY] When used to represent the earth, an oblate ellipsoid of revolution, made by rotating an ellipse about its minor axis. *See also* spheroid.

ellipsoid

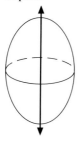

ellipticity *See* eccentricity.

envelope The rectangle defined by one or more geographical features in coordinate space, determined by the minimum and maximum coordinates in the x and y directions, as well as the ranges of any z- or m-values that the features may have.

ephemeris [ASTRONOMY, NAVIGATION, GPS] A list of the predicted positions of a satellite for each day of the year, or for other regular intervals.

equal area classification Classifies polygon features so that the total area of polygons in each class is approximately the same.

equal area projection A projection in which the whole of the map as well as each part has the same proportional area as the corresponding part on earth. An equal area projection may distort shape, angle, scale, or any combination thereof. No flat map can be both equal area and conformal.

equal interval classification Divides the range of attribute values of a set of geographic features by the number of classes. Classes break at equal intervals, regardless of how many members they contain.

equator The parallel of reference that is equidistant from the poles and defines the origin of latitude values as 0 degrees north or south.

equator

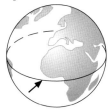

equatorial aspect A planar (or azimuthal) projection with its center located at the equator.

equiangular *See* conformality.

equiareal projection *See* equal area projection.

equidistant projection A projection that maintains scale along one or more lines, or from one or two points to all other points on the map. Lines along which scale (distance) is correct are the same proportional length as the lines they reference on the globe. In the sinusoidal projection, for example, the central meridian and all parallels are their true lengths. An azimuthal equidistant projection centered on Chicago shows the correct distance between Chicago and any other point on the projection, but not between any other two points. No flat map can be both equidistant and equal area.

equivalent projection *See* equal area projection.

Erase [GEOPROCESSING] An ArcInfo command that removes features from one coverage that overlap features in another coverage. *Compare* Clip.

Euclidean distance The straight-line distance between two points, normally on a plane. The distance can be calculated using the Pythagorean theorem.

event A geographic location stored in tabular rather than spatial form. *See also* dynamic segmentation, route event.

executable file A binary file containing a program in machine language that can be executed (run).

export 1. To move data from one computer system to another, and often, in the process, from one file format to another. 2. An ArcInfo command that creates an interchange file, or E00 file, for transferring coverages between different systems.

E

exposure station [REMOTE SENSING] Also **camera station, air station** In aerial photography, each point in the flight path at which the camera exposes the film.

external polygon *See* universe polygon.

extrusion Projecting features in a data layer into three-dimensional space. Points become vertical lines, lines become walls, and polygons become solid blocks.

false easting The value added to all x-coordinates of a map projection so that none of the values in the geographic region being mapped are negative.

false northing The value added to all y-coordinates of a map projection so that none of the values in the geographic region being mapped are negative.

feature 1. An object in a landscape or on a map. 2. A shape in a spatial data layer, such as a point, line, or polygon, that represents a geographic object.

feature attribute table (FAT) A table that stores the attribute information for a specific feature class in a coverage. Types of feature attribute tables include the PAT (for polygons or points), AAT (for arcs), NAT (for nodes), RAT (for routes), SEC (for sections), and TAT (for annotation).

feature class In a shapefile, coverage, or geodatabase, a collection of spatial data with the same shape type (e.g., point, line, or polygon).

feature data set In a geodatabase, a collection of feature classes that share the same spatial reference.

field 1. Also **item** A vertical column in a table that represents some characteristic for all of the records in the table, given in numbers or words. 2. The place in a database record, or in a graphical user interface, where data can be entered. 3. A synonym for surface.

file Information stored on disk or tape. A file may be a collection of data, a document (text file), or a program (executable file). It generally resides within a directory, and always has a unique name.

file header Also **header file** The part of a file—usually the first part—that contains metadata, or information about the file itself.

file name The name that distinguishes a file from all other files in a particular directory. It can refer to the name of the file by itself (harold), the name plus the file extension (harold.shp), or the whole path of a file up to and including the file name extension (C:\mygisdata\shapefiles\harold.shp).

F

file name extension Also **file extension** The abbreviation following the final period in a file name that indicates the file's format, for example, cities.shp, soils.zip, pigeons.tif. It is usually one to three letters long.

filter 1. Any device that separates desired information from undesired information. 2. [REMOTE SENSING, ARC GRID, DIGITAL IMAGE PROCESSING] A matrix of numbers used to mathematically modify pixel or grid cell values.

fix A single position obtained by surveying, GPS, or astronomical measurements, usually given with altitude, time, date, and latitude/longitude or grid position.

flatbed scanner A scanner in which a map or image is placed on a flat surface and is converted to digital format by a sensor beam that moves across it. *Compare* drum scanner.

flattening Also **polar flattening, ellipticity, eccentricity** A measure of how much a spheroid differs from a sphere. The flattening is the ratio of the semimajor axis minus the semiminor axis to the semimajor axis.

flow map A map that uses line symbols of different thickness to show the proportion of traffic or flow within a network.

folder *See* directory.

font In traditional typesetting, the complete set of characters of one size (14 point) of one typeface (Centaur bold italic) of a particular type family (Centaur). In digital typesetting, font is commonly used to mean typeface.

foreign key An item (column) in a table that can uniquely identify records in another table. A foreign key in one table is the primary key of another, related table. The link between the two defines a relational join. *See also* primary key.

form lines Lines on a map that resemble contour lines but do not refer the shape of terrain to a true datum and do not use regular spacing.

forms interface *See* GUI.

fractal A geometric shape that repeats itself, at least roughly, at all scales. Examples of fractals include the Koch snowflake, the Mandelbrot set, and the Lorenz attractor. Fractals can be used to model complex natural shapes such as clouds and coastlines.

frequency Of a wave of energy, the number of oscillations per unit of time, or the number of wavelengths that pass a point in a given amount of time.

from-node Of an arc's two endpoints, the one first digitized.

fuzzy tolerance The minimum distance separating all arc coordinates (nodes and vertices) in a coverage, within which two points will be treated as one. Fuzzy tolerance also defines the distance that a coordinate can move during certain operations, such as Clean. It is a very small distance, usually from 1/10,000 to 1/1,000,000 times the width of the coverage extent, and is generally used to correct inexact intersections. *See also* snapping.

gazetteer A list of geographic places and their coordinates, along with other information such as area, population, and cultural statistics.

GBF/DIME (geographic base files/Dual Independent Map Encoding) Vector geographic base files made for the 1970 and 1980 censuses, containing address ranges, ZIP Codes, and the coordinates of street segments and intersections for most metropolitan areas. DIME was replaced by TIGER for the 1990 census.

GDOP Geometric Dilution of Precision. *See* Dilution of Precision.

generalization 1. Reducing the number of points in a line without losing its essential shape. 2. Enlarging and resampling cells in a raster format. 3. [CARTOGRAPHY] Any reduction of information so that a map is clear and uncluttered when its scale is reduced.

geocentric [ASTRONOMY, GEODESY] 1. Having the earth as a center. 2. Measured from the earth or the earth's center.

geocentric coordinate system A three-dimensional coordinate system with its origin at or near the center of the earth and with three mutually perpendicular axes. The z-axis lies in the earth's axis of rotation. The x-axis is in the plane of the equator, and passes through the Greenwich meridian. The y-axis also lies in the plane of the equator, forming a right-handed coordinate system.

geocentric datum Also **earth-centered datum, geocentric geodetic datum** [GEODESY] A datum in which the center of the ellipsoid is either located at or related to the earth's center of mass.

geocentric latitude [GEODESY] The angle between the equatorial plane and a line from a point on the surface to the center of the sphere or spheroid. On a sphere, all latitudes are geocentric. The unqualified term *latitude* generally refers to geographic, or geodetic, latitude. *Compare* geodetic latitude.

geocentric longitude The angle between the prime meridian and a line drawn from a point on the surface to the center of the earth. For an ellipsoid of revolution (such as the earth), geocentric longitude is the same as geodetic longitude.

geocode A code representing the location of an object, such as an address, a census tract, a postal code, or x,y coordinates.

geocoding *See* address geocoding.

geodatabase An ArcInfo 8 data storage format. A geodatabase represents geographic features and attributes as objects and is hosted inside a relational database management system.

geodesic 1. The shortest distance between two points on the surface of a spheroid. Any two points along a meridian form a geodesic. 2. *See* geodetic.

geodesy The science that determines the size and shape of the earth and measures its gravitational and magnetic fields.

geodetic Also **geodesic** Pertaining to geodesy; relating to the geometry of the earth's surface or to curved surfaces in general.

geodetic datum A datum designed to best fit all or part of the geoid, defined by the origin of an initial point (its latitude, longitude, and according to some authors, its elevation); the orientation of the network (the azimuth of a line from the origin); and two constants that define the reference spheroid. More recent definitions express the position and orientation of the datum as functions of the deviations in the meridian and in the prime vertical, the geoid-ellipsoid separation, and the parameters of a reference ellipsoid.

geodetic latitude The angle that a line drawn perpendicular to the surface through a point on a spheroid makes with the equatorial plane.

geodetic longitude The angle between the plane of the meridian that passes through a point on the surface of the spheroid and the plane of an arbitrarily chosen initial meridian, usually Greenwich.

geodetic reference system Also **geographic reference system** *See* geodetic datum.

Geodetic Reference System of 1980 (GRS80) The standard measurements of the earth's shape and size adopted by the International Union of Geodesy and Geophysics in 1979.

geodetic survey A survey that takes the figure and size of the earth into account, used to precisely locate horizontal and vertical positions suitable for controlling other surveys.

G

geographic base file (GBF) A database of files containing cartographic and attribute information such as boundaries of geographic areas, address ranges, and street intersections. The most common GBFs are DIME files and TIGER files.

geographic coordinates Locations on the surface of the earth expressed in degrees of latitude and longitude.

geographic coordinate system [GEODESY, NAVIGATION, SURVEYING] A reference system using latitude and longitude to define the locations of points on the surface of a sphere or spheroid.

geographic data Information about geographic features, including their locations, shapes, and descriptions.

geographic database A collection of spatial data and its attributes, organized for efficient storage and retrieval.

geographic feature *See* feature.

geographic grid *See* graticule.

geographic information system (GIS) A collection of computer hardware, software, and geographic data for capturing, storing, updating, manipulating, analyzing, and displaying all forms of geographically referenced information.

geographic latitude *See* geodetic latitude.

geographic longitude *See* geodetic longitude.

geographic north Also **true north** The direction from any point on the earth's surface to the north geographic pole. *Compare* magnetic north.

geography 1. The study of the earth's surface, especially how climate and elevation interact with soil, vegetation, and animal populations. 2. The geographic features of an area. 3. A word game in which each player in rotation says aloud a geographic place name beginning with the last letter of the place name mentioned by the preceding player.

geoid [GEODESY] The exact figure of the earth considered as a mean sea level extended continuously through the continents. The geoid varies from the ellipsoid model by as much as 80 meters above and 60 meters below its surface.

geoid

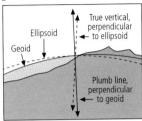

geoid–ellipsoid separation The distance from the surface of an ellipsoid to the surface of the geoid, measured along a line perpendicular to the ellipsoid. The separation is positive if the geoid lies above the ellipsoid, negative if it lies below.

geoid height 1. The height of the geoid above the ellipsoid in use (usually the WGS84 ellipsoid). 2. The height of a point above the geoid, often called elevation above mean sea level.

geometric correction [REMOTE SENSING, PHOTOGRAMMETRY] The correction of errors in remotely sensed data caused by satellites not staying at a constant altitude or by sensors deviating from the primary focal plane. The images are compared to ground control points on accurate basemaps and resampled, so that exact locations and appropriate values for pixel brightness can be calculated.

Geometric Dilution of Precision *See* Dilution of Precision.

geometric network A one-dimensional nonplanar graph, composed of topologically connected edge and junction features, that represents a linear network such as a road, utility, or hydrologic system.

geoprocessing GIS operations such as geographic feature overlay, coverage selection and analysis, topology processing, and data conversion.

georeference To assign coordinates from a known reference system, such as latitude/longitude, UTM, or State Plane, to the page coordinates of an image or a planar map.

georelational data model The data model used in ArcInfo coverages that represents geographic features as an interrelated set of spatial and tabular data.

geostationary satellite A satellite positioned approximately 35,790 kilometers above the earth's equator, with an inclination and an eccentricity approaching zero. At this height it orbits as fast as the earth rotates on its axis, so it remains effectively stationary above a point on the equator. A geostationary satellite is geosynchronous, but the reverse is not necessarily the case. *See also* Clarke Belt.

geosynchronous satellite A satellite moving west to east whose orbital period is equal to the earth's rotational period. If the orbit is circular and lies in the plane of the equator, the satellite will remain over one point on the equator and is termed geostationary. If not, the satellite will appear to make a figure eight once a day between the latitudes that correspond to its angle of inclination over the equator.

Global Navigation Satellite System (GLONASS) The Russian counterpart to the United States' GPS. *See* Global Positioning System.

Global Positioning System (GPS) A constellation of twenty-four satellites, developed by the U.S. Department of Defense, that orbit the earth at an altitude of 20,200 kilometers. These satellites transmit signals that allow a GPS receiver anywhere on earth to calculate its own location. The Global Positioning System is used in navigation, mapping, surveying, and other applications where precise positioning is necessary.

gnomonic projection A planar projection, tangent to the earth at one point, that views the earth's surface from the center of the globe. Used by Thales to chart the heavens, it is possibly the oldest map projection.

gore A crescent-shaped map of an area that lies between two lines of longitude. A gore can be fitted to the surface of a globe with very little distortion.

GPS *See* Global Positioning System.

gradient 1. Also **slope, grade** The ratio between vertical distance (rise) and horizontal distance (run), often expressed as a percentage. A 10-percent gradient rises 10 feet for every 100 feet of horizontal distance. 1a. An inclined surface. 2. [PHYSICS] The rate at which a quantity such as temperature or pressure changes in value.

graduated color map A map that uses a range of colors to indicate a progression of numeric values. For example, differences in population density could be represented by increasing the saturation of a single color, and temperature changes could be represented by colors ranging from blue to red.

graduated symbol map A map whose symbols correspond in size to the amount of the attribute they represent. For example, larger rivers could be represented by thicker lines, and denser populations by larger dots.

grain tolerance A parameter controlling the number of vertices and the distance between them on lines that represent curves. The smaller the grain tolerance, the closer the vertices can be. Unlike densify tolerance, grain tolerance can affect the shape of curves. *See* spline, *compare* densify.

granularity [DIGITAL IMAGE PROCESSING, PHOTOGRAMMETRY] The objective measure, using a microdensitometer, of the random groupings of grains into denser and less dense areas in a photographic image.

graphical user interface *See* GUI.

graphic database A collection of digital map features (such as points, lines, polygons, or pixels) and annotations that can be used to generate a display. *Compare* geographic database.

graphic elements [CARTOGRAPHY] The basic characteristics of any map symbol: size, position, shape, spacing, hue, value, saturation, brightness, orientation, and pattern.

graphic scale *See* bar scale.

graphics page The area on a graphics display device reserved for map display, or for simulating the plotter page area. Page units are typically in centimeters or inches instead of ground coordinates such as meters or feet.

graphics tablet A small digitizer, usually about 11 inches square, used for interactive digitizing, although not generally at the same level of precision as a full-sized digitizing table.

graticule 1. [MAPPING, GEODESY] A network of longitude and latitude lines on a map or chart that relates points on a map to their true locations on the earth. 2. [ASTRONOMY] A glass plate or cell with a grid or cross wires on it that rests in the focal plane of the eyepiece of a telescope, used to locate and measure celestial objects.

graticule

G

gravimeter [GEODESY] A weight on a spring that is pulled downward where gravity is stronger, used to measure small variations in the earth's gravitational field.

gravimetric geodesy The science of deducing the size and shape of the earth by measuring its gravitational field.

gravity modeling [GEOGRAPHY, ENGINEERING] An approach to modeling population that assumes that the influence of populations on one another varies inversely with the distance between them.

grayscale 1. All the shades of gray from white to black. 2. Levels of brightness for displaying information on a monochrome display device.

great circle [NAVIGATION, GEODESY] A circle or near circle produced by the intersection of a sphere and a flat plane that passes through the center of the sphere. The equator and all lines of longitude are great circles.

great circle route The shortest distance between two points on a sphere.

Greenwich mean time (GMT) Also **Coordinated Universal Time (UTC), Universal time (UT)** The mean solar time on the 0-degree meridian at Greenwich, reckoned from midnight. Greenwich mean time is the basis for standard time worldwide.

Greenwich meridian Also **prime meridian, international meridian** The meridian adopted by international agreement in 1884 as the 0-degree meridian from which all other longitudes are calculated.

grid 1. Equally sized square cells arranged in rows and columns. Each cell contains a value for the feature it covers. *See also* raster. 2. [CARTOGRAPHY] Any network of parallel and perpendicular lines superimposed on a map, usually named after the map's projection, such as a Lambert grid, or transverse Mercator grid.

grid cell 1. A single square in a grid that represents a portion of the earth, such as a square meter or square mile. Each grid cell has a value for the feature or attribute that it covers, such as soil type, census tract, or vegetation class. 2. A pixel.

grid lines *See* graticule.

grid reference system A reference system that uses a rectangular grid to assign x,y coordinates to individual locations. *See* Cartesian coordinate system.

ground control Also **control mapping** [SURVEYING, REMOTE SENSING, PHOTOGRAMMETRY] A system of points with established positions, elevations, or both, used as fixed references in relating map features, aerial photographs, or remotely sensed images.

ground control point [SURVEYING, REMOTE SENSING, PHOTOGRAMMETRY] Also **control point, control station** A point on the ground whose location has been determined by a horizontal coordinate system or a vertical datum.

ground receiving station Communications equipment for receiving and transmitting signals to and from satellites such as Landsat.

ground truth *See* ground control.

GUI (graphical user interface) Pronounced "gooey." [COMPUTING] A program interface in which the user clicks on graphic icons and menus with a mouse instead of typing commands with the keyboard. *Compare* command-line interface.

hachured contour On a topographic map, concentric contour lines drawn with hachures to indicate a closed depression, or basin. Concentric contour lines drawn without hachure marks indicate a hill.

hachures [CARTOGRAPHY] Lines on a map that indicate the direction and steepness of slopes. For steep slopes the lines are short and close together; for gentle slopes they are longer, lighter, and farther apart. Contours, shading, and hypsometric tints have largely replaced hachuring on modern maps.

halftone image A continuous tone image photographed through a fine screen that converts it into uniformly spaced dots of varying size while maintaining all the gradations of highlight and shadow. The size of the dots varies in proportion to the intensity of the light passing through them.

Hamiltonian circuit A path through a network that visits each stop in the network only once and then returns to its point of origin.

Hamiltonian path A path through a network that visits each stop in the network once without returning to its point of origin.

HDOP Horizontal Dilution of Precision. *See* Dilution of Precision.

heading [NAVIGATION] The direction of a moving object from a point of observation, expressed as an angle from a known direction, usually north. Bearing and heading differ in that bearing refers to a fixed position, whereas heading refers to the direction in which an object is moving.

hierarchical database A database that stores related information in a structure very like a tree, where records can be traced to parent records which in turn can be traced to a root record.

high-pass filter [DIGITAL IMAGE PROCESSING] A spatial filter that blocks low-frequency (long-wave) radiation, resulting in a sharpened image. *See* edge enhancement.

hillshading Also **relief shading** 1. [CARTOGRAPHY] Shadows drawn on a map to simulate the effect of the sun's rays over the land. 2. On a grid, the same effect achieved by assigning an illumination value from 0 to 255 to each cell according to a specified azimuth and altitude for the sun.

hillshading

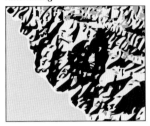

histogram A graph showing the distribution of values in a set of data. Individual values are displayed along a horizontal axis, and the frequency of their occurrence is displayed along a vertical axis.

histogram equalization [DIGITAL IMAGE PROCESSING] The redistribution of pixel values so that each range contains approximately the same number of pixels. The resulting histogram is nearly flat.

horizon [ASTRONOMY, GEODESY, SURVEYING] 1. Also **apparent, visible, local, sensible,** or **topocentric horizon** The apparent or visible junction of land and sky. 2. Also **true horizon** The horizontal plane tangent to the earth's surface and perpendicular to the line through an observer's position and the zenith of that position. The apparent or visible horizon approximates the true horizon only when the point of vision is very close to sea level. 3. Also **astronomical horizon** The great circle in which an observer's horizon meets the celestial sphere. 4. [CARTOGRAPHY] The edge of a map projection.

horizon circle The circle containing all points equidistant from the center of a zenithal projection.

horizontal control [GEODESY, SURVEYING] A network of known horizontal geographic positions, referenced to the geographic parallels and meridians or to other lines of orientation such as plane coordinate axes.

horizontal control datum Also **horizontal datum, horizontal geodetic datum** [GEODESY, SURVEYING] A geodetic reference point that is the basis for horizontal control surveys and consists of five quantities: latitude, longitude, the azimuth of a line from the reference point, and two constants that are the parameters of the reference ellipsoid. The datum may extend over an area of any size.

hot link A link that connects a geographic feature to an external image, text, or executable file. When the feature is clicked, the file runs or is displayed on-screen.

hub [NETWORK ANALYSIS] A central node in a network for routing goods to their destinations.

hue The dominant wavelength of a color, by which it can be distinguished as red, green, yellow, blue, and so forth.

hydrographic datum Also **chart datum** A plane of reference for depths, depth contours, and elevations of foreshore and offshore features.

hydrographic survey [OCEANOGRAPHY, GEODESY, NAVIGATION] Survey of a water body, particularly of its currents, depth, submarine relief, and adjacent land.

hydrologic cycle The movement of the earth's free water from the oceans through the atmosphere to the land and back again.

hydrology The science that deals with the properties and distribution of the waters of the earth.

hypsography 1. The study of the earth's topography above sea level, especially the measurement and mapping of land elevation. 2. Relief features on a map.

hypsometric map A map showing relief, whether by contours, hachures, shading, or tinting.

hypsometric tinting Also **layer tinting, altitude tinting** Relief or depth depicted by a gradation of colors, usually between contour lines. Each color represents a different elevation.

hypsometry 1. The science that determines the distribution of elevations above an established datum, usually sea level. 2. The determination of terrain relief, by any method. 3. Relief features on a map.

Identity [GEOPROCESSING] An overlay that computes the geometric intersection of two coverages. The output coverage preserves all the features of the first coverage plus those portions of the second (polygon) coverage that overlap the first. For example, a road passing through two counties would be split into two arc features, each with the attributes of the road and the county it passes through. *Compare* Intersect, Union.

image 1. A graphic representation of a scene, typically produced by an optical or electronic device such as a camera or a scanning radiometer. 2. [REMOTE SENSING] A graphic representation of a scene stored as a raster of pixels, each of which has a numeric value that represents the intensity of reflected light, heat, or other electromagnetic radiation for the specific area that it covers. The term is generally used when the radiation is not recorded directly on film. 3. [COMPUTING, GRAPHICS] A description of a picture, stored either as a set of brightness and color pixel values (a bit map) or as a set of instructions for drawing the image (a metafile).

image catalog A set of images that are geographically referenced and can be accessed as one image.

image division [DIGITAL IMAGE PROCESSING] Dividing the pixel values in an image by the values of the corresponding pixels in a second image to increase the contrast between features. Normally used for identifying concentrations of vegetation.

image pair *See* stereopair.

image processing *See* digital image processing.

imager Any satellite instrument that measures and maps the earth and its atmosphere.

image scale [REMOTE SENSING] The ratio between a distance in a photograph and the actual distance on the ground, calculated as focal length divided by the flying height above mean ground elevation. Image scale can vary in a single image from point to point due to surface relief and the tilt of the camera lens.

impedance [NETWORK ANALYSIS] The amount of resistance, or cost, required to traverse a line from its beginning to its end, or to make a turn from one line, through a node, onto another line. Impedance may be a measure of travel distance, time, speed of travel multiplied by distance, and so on. An optimum path in a network is the path of least resistance (or lowest impedance).

impedance model A routing model that determines the best route by finding the path of least resistance.

import To load data from one computer system or application into another. Importing often involves some form of data conversion.

incident energy Electromagnetic radiation striking a surface.

index A data structure used to speed the search for records in a database or for spatial features in geographic data sets. In general, unique identifiers stored in a key field point to records or files holding more detailed information.

index contour line On a topographic map, a contour line that is heavier than the rest and usually labeled with the elevation or depth that it represents. Every fourth or fifth contour line may be an index line, depending on the contour interval.

index map Also **key map** A schematic map used as a reference for a collection of map sheets, outlining the total area covered and usually providing a reference code for each map.

INFO A tabular DBMS used to store and manipulate feature attribute tables. For each set of coverages in a workspace, a set of INFO data files, feature attribute tables, and related files are stored in a subdirectory, also called INFO.

information system A system that contains or is related to a database of information and also provides the means of data storage, retrieval, and analysis, so that a user may query and receive answers from the database.

infrared scanner Also **thermal mapper** A device that detects infrared radiation and converts it into an electrical signal that can be recorded on film or magnetic tape.

infrastructure The system of roads, bridges, canals, cables, wires, pipes, reservoirs, and sewers that provide public services to an area.

inset map A small map set within a larger map. An inset map might show an area that does not fit neatly into the main map, or a detail of part of the map at a larger scale, or the context of the area covered by the map at a smaller scale.

inset map

intensity In the IHS (intensity, hue, saturation) color model, brightness ranging from black to white.

interface For the purpose of data communication, a hardware and software link that connects two computer systems, two applications, a computer and its peripherals, or a computer and its user. *See* graphical user interface, command-line interface.

international date line A meridian of longitude lying 180 degrees east and west of the Greenwich meridian, dividing the world's time zones into those that are twelve hours ahead or twelve hours behind Greenwich standard time. A traveler going west across the date line adds a day; a traveler going east across it subtracts a day.

international date line

interpolation Estimating an unknown value that falls between known values.

interrupted projection A world projection that reduces distortion by dividing the projected area into gores, each with its own central meridian.

53

Intersect [GEOPROCESSING] An overlay that preserves the features and attributes that fall within the area common to two coverages. *Compare* Identity, Union.

Intersect

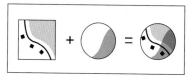

inverse distance weighted (IDW) An interpolation technique that determines cell values in a grid or image with a set of sample points that have been weighted so that the farther a point is from the cell being evaluated, the less important it is in calculating the cell's value.

isarithm Also **isoline** [CARTOGRAPHY] 1. A line connecting points on a surface of equal value. 2. A map of such lines. *See also* isometric line, isopleth.

isoline map

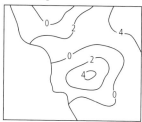

island polygon A polygon enclosed by another, larger polygon.

isobar A line on a weather map connecting places of equal barometric pressure.

isogonic line A line on a map or chart that connects points of equal magnetic variation.

isoline *See* isarithm.

isometric line [CARTOGRAPHY] An isarithm drawn according to known values, either sampled or derived, that can occur at points. Examples of sampled quantities that can occur at points are elevation above sea level, an actual temperature, or an actual depth of precipitation. Examples of derived values that can occur at points are the average of temperature over time for one point or the ratio of smoggy days to clear days for one point. *Compare* isopleth.

isopleth Also **isoplethic line** [CARTOGRAPHY] An isarithm drawn according to known values that can only be recorded for areas, not points. Examples include population per square mile or the ratio of residential land to total land for an area. *Compare* isometric line.

isotherm A line on a map connecting points of equal temperature.

item Also **field** A column in an attribute table that contains the values of one attribute for each record in the table. *See also* record, attribute table.

iterative procedure A procedure that is repeated over and over again.

jaggies Also **aliasing** [GRAPHICS] The sawtooth effect observable when a curve is drawn on a raster display.

join Appending the fields of one table to those of another through a common item. A join is usually used to attach more attributes to the attribute table of a geographic layer. *See also* relational join, spatial join.

junction 1. A node joining two or more arcs. 2. In a linear network, a feature that occurs at the intersection of two or more edges and allows flow between them.

$\mathcal{K}$

key Also **key attribute, primary key** A column in a database that stores a unique value for each record. *See also* foreign key.

key map *See* index map.

kinematic positioning [GPS] Determining the position of an antenna on a moving object such as a ship or an automobile. *Compare* static positioning.

kriging An interpolation technique that assumes that the spatial variation in the data being modeled is statistically homogeneous throughout the surface. It is often used to estimate surface elevations from known elevations at specific points.

label Text placed next to a feature on a map to describe or identify it.

labels

label point A coverage feature class used to represent point features or identify polygons. When representing point features, the x,y location of the point describes the location of the feature. When identifying polygons, the point can be located anywhere within the polygon.

LAN *See* local area network.

landform Any natural feature of the land having a characteristic shape, including major forms such as plains and mountains and minor forms such as hills and valleys.

land information system (LIS) A geographic information system for cadastral and land-use mapping, typically used by local governments.

landmark 1. [SURVEYING, NAVIGATION] Any prominent natural or artificial object in a landscape used to determine distance, bearing, or location. 2. A building or location that has historical or architectural value.

land parcel An area of land for which rights of ownership and use can be bought.

Landsat [REMOTE SENSING] Earth-orbiting satellites developed by NASA that gather imagery for land-use inventory, geological and mineralogical exploration, crop and forestry assessment, and cartography.

land use The classification of land according to how it is used; for example, agricultural, industrial, residential, urban, rural, or commercial. Natural features of the land such as forest, pastureland, brushland, and bodies of water are also often classified in this manner.

large scale [CARTOGRAPHY] Generally, a map scale whose representative fraction is 1:50,000 or larger. A large-scale map shows a small area on the ground at a high level of detail. *See* scale, representative fraction.

latitude [NAVIGATION, GEODESY] The angular distance along a meridian north or south of the equator, usually measured in degrees. Lines of latitude are also called parallels. *See also* geodetic latitude and geocentric latitude.

latitude–longitude Also **lat/long, lat/lon** [NAVIGATION, GEODESY] The most commonly used spherical reference system for locating positions on the earth. Latitude and longitude are angles measured from the equator and the prime meridian to locations on the earth's surface. Latitude measures angles in a north–south direction; longitude measures angles in the east–west direction.

latitude of center The latitude value that defines the center, and sometimes the origin, of a projection.

latitude of origin The latitude value that defines the origin of the y-coordinate values for a projection.

lattice A rectangular array of points spaced evenly in the x and y directions from a common origin. Lattices and grids are stored in the same manner, with each lattice point corresponding to the center of a cell in an equivalent grid. In a grid, a location lying anywhere within a cell is given the attribute value of that cell. In a lattice, a location will only receive the attribute value of a lattice point if it lies directly on that point. The attribute value of a location not directly on a lattice point is interpolated from the values of the lattice points surrounding it. *See also* grid.

layer 1. A set of vector data organized by subject matter, such as roads, rivers, or political boundaries. Vector layers act as digital transparencies that can be laid atop one another for viewing or spatial analysis. 2. A set of raster data representing a particular geographic area, such as an aerial photograph or a remotely sensed image. In both (1) and (2), layers covering the same geographical space are registered to one another by means of a common coordinate system. 3. A file that stores symbology and display information for a given vector or raster data set. The layer does not actually contain the data, but points to its physical location.

layout [CARTOGRAPHY] 1. The way map elements such as the title, legend, and scale bar are arranged on a printed map. 2. An on-screen document where said map elements are arranged for printing.

L-band The group of radio frequencies that carry data from GPS satellites to GPS receivers.

least-cost path [NETWORK ANALYSIS] The path between two points on a network that costs the least to traverse, where cost is a function of time, distance, or some other factor defined by the user. *See also* impedance.

left–right topology The data structure in an ArcInfo coverage that stores, for each arc, the identity of the polygons to the left and right of it. *See also* topology.

legend [CARTOGRAPHY] The reference area on a map that lists and explains the colors, symbols, line patterns, shadings, and annotation used on the map, and often includes the map's scale, origin, and projection.

legend

leveling [SURVEYING] The measure of the heights of objects and points according to a specified elevation, usually mean sea level.

lidar [REMOTE SENSING] Short for light intensity detection and ranging. Lidar uses lasers to measure distances to reflective surfaces.

line Also **linear feature** A shape having length and direction but no area, connecting at least two x,y coordinates. Lines represent geographic features too narrow to be displayed as an area at a given scale, such as contours, street centerlines, or streams, or linear features with no area, such as state and county boundary lines.

linear scale *See* bar scale.

linear units The unit of measure in a planar coordinate system, often meters or feet. Map projection parameters such as false easting and false northing are defined in linear units.

line chart A chart in which data points are displayed against x and y axes and are connected (sometimes approximately) by a single line. Line charts are good for representing trends among data values over a period of time.

line-in-polygon [GEOPROCESSING] A spatial operation in which lines in one geographic data layer are overlaid with the polygons of another to determine which lines are contained within the polygons. The lines in the resulting data layer receive the attributes of the polygons that contain them.

line of sight 1. A line drawn between two points, an origin and a target, in a three-dimensional scene that shows whether the target is visible from the origin and, if it is not visible, where the view is obstructed. 2. In a perspective view, the point and direction from which the viewer looks into the image.

line smoothing Adding extra points to lines to reduce the sharpness of angles between line segments, resulting in a smoother appearance. *Compare* weeding.

line symbol A cartographic symbol type for representing borders, neatlines, boundaries, rivers, streets, and so forth.

line thinning *See* weeding.

link 1. A coverage feature class used in rubber sheeting. A link is a line whose endpoints represent the from- and to-locations of a point to be moved. 2. An operation that relates two tables using a common field, without altering either table. *Compare* join.

LIS *See* land information system.

local area network (LAN) Communications hardware and software that connects computers in a small area such as a room or a building. Computers in a LAN can share data and peripheral devices such as printers and plotters, but have no necessary link to outside computers. *Compare* WAN.

location Also **position** A point on the earth's surface or in geographical space described by x-, y-, and z-coordinates, or by other precise information such as a street address.

location–allocation Finding the best locations for one or more facilities that will service a given set of points and then assigning those points to the facilities, taking into account factors such as the number of facilities available, their cost, and the maximum impedance from a facility to a point.

L

locking In a shared database, a mechanism that only allows one person at a time to edit a file. Other users may view it, but generally only the first person to retrieve the file is able to alter it.

log file A history file, usually text, containing a list of the commands used to perform a function or procedure.

logical expression *See* Boolean expression.

logical operator *See* Boolean operator.

logical selection Also **logical query** Using Boolean expressions to select features from a geographic layer based on their attributes; for example, "select all polygons with an area greater than 16,000 units" or "select all street segments named Green Apple Run."

longitude The angular distance, expressed in degrees, minutes, and seconds, of a point on the earth's surface east or west of a prime meridian (usually the Greenwich meridian). All lines of longitude are great circles that intersect the equator and pass through the north and south poles.

longitude of center The longitude value that defines the center, and sometimes the origin, of a projection.

longitude of origin The longitude value that defines the origin of the x-coordinate values for a projection.

lookup table A tabular data file that contains additional attributes for records stored in an attribute table.

low-pass filter [DIGITAL IMAGE PROCESSING] A spatial filter that blocks high-frequency (shortwave) radiation, resulting in a smoother image.

loxodrome *See* rhumb line.

macro A file, usually text, containing a sequence of commands that are executed as one command. Macros are used to perform repetitive or complicated operations.

magnetic declination *See* declination.

magnetic north Also **compass north** The direction from a point on the earth's surface following a great circle toward the magnetic north pole, indicated by the north-seeking end of a compass.

magnetometer An instrument used to measure variations in the strength and direction of the earth's magnetic field.

major axis The longer diameter of an ellipse or spheroid.

map 1. A graphic depiction on a flat surface of the physical features of the whole or a part of the earth or other body, or of the heavens, using shapes or photographic imagery to represent objects, and symbols to describe their nature; at a scale whose representative fraction is less than 1:1, generally using a specified projection and indicating the direction of orientation. 2. Any graphical presentation of geographic or spatial information.

map generalization Decreasing the level of detail on a map so that it remains uncluttered when its scale is reduced.

map library A collection of geographic data partitioned spatially as a set of tiles and thematically as a set of layers, indexed by location for rapid access.

map projection [CARTOGRAPHY] A mathematical model that transforms the locations of features on the earth's curved surface to locations on a two-dimensional surface. It can be visualized as a transparent globe with a lightbulb at its center casting lines of latitude and longitude onto a sheet of paper. Generally, the paper is either flat and placed tangent to the globe (a planar or azimuthal projection), or formed into a cone or cylinder and placed over the globe (cylindrical and conical projections). Every map projection distorts distance, area, shape, direction, or some combination thereof.

M

map query Asking spatial or logical questions of the data in a GIS. A spatial query selects features on the basis of their location or spatial relationship to each other. A logical query selects features whose attributes meet specific criteria; for example, all polygons whose value for AREA is greater than 10,000, or all arcs whose value for NAME is "Main St."

map series A collection of maps covering a limited region and addressing a particular theme, using a common scale and projection.

map units The ground units in which the coordinates of spatial data are stored, such as feet, miles, meters, or kilometers.

marker symbol A symbol used to represent a point location on a map.

mask 1. A grid theme that excludes areas on another grid theme from analysis. 2. [GRAPHICS] A grayscale image that excludes areas on another image from manipulation or display.

mass points Irregularly distributed sample points, each with an x-, y-, and z-value, used to build a TIN. Ideally, mass points are chosen so as to capture the more important variations in the shape of the surface being modeled.

mean The average of a set of values, calculated by dividing the sum of the values by the number of values.

mean sea level The average height of the surface of the sea for all stages of the tide over a nineteen-year period, usually determined by averaging hourly height readings from a fixed level of reference.

measure A value stored along a linear feature that represents a location relative to the beginning of the feature, or some point along it, rather than as an x,y coordinate. Measures are used to map events such as distance, time, or addresses along linear features. *See also* route, dynamic segmentation.

median The middle value of a set of values when they are ordered by rank.

medium scale Generally, a map scale whose representative fraction is between 1:50,000 and 1:500,000. *See* scale, representative fraction.

mereing Also **mering** [SURVEYING] Establishing a boundary relative to ground features present at the time of the survey.

merge *See* Dissolve, Union.

meridian [NAVIGATION, GEODESY] A great circle on the earth that passes through the poles, often used synonymously with longitude. From a prime meridian or 0 degrees longitude (usually the meridian that runs through the Royal Observatory in Greenwich, England), measures of longitude are negative to the west and positive to the east, where they meet halfway around the globe at the line of 180 degrees longitude.

metadata Information about a data set. Metadata for geographical data may include the source of the data; its creation date and format; its projection, scale, resolution, and accuracy; and its reliability with regard to some standard.

metes and bounds [SURVEYING] The limits of a land parcel identified as relative distances and bearings from natural or human-made landmarks. Metes and bounds surveying is often used for areas that are irregularly shaped.

microdensitometer A densitometer that can read densities in minute areas, used particularly for studying spectroscopic and astronomical images.

micrometer 1. Also **micron** Symbol μm. One millionth of a meter; used to measure wavelengths in the electromagnetic spectrum. 2. [ASTRONOMY, ENGINEERING] An instrument for measuring minute lengths or angles.

minimum mapping units For a given map scale, the size in map units below which a long narrow feature will be represented by a line and a small area by a point.

minor axis The shorter diameter of an ellipse or spheroid.

minute 1. Also **angular minute, minute of arc** An angle equal to one sixtieth of a degree of latitude or longitude and containing sixty seconds. 2. A unit of time equal to one sixtieth of an hour and containing sixty seconds.

mixed pixel Also **mixel** [REMOTE SENSING] A pixel whose digital number represents the average of several spectral classes within the area that it covers on the ground, each emitted or reflected by a different type of material. Mixed pixels are common along the edges of features.

model 1. An abstraction of reality. 2. A set of rules and procedures for representing a phenomenon or predicting an outcome. The terms modeling and analysis are often interchanged, although modeling implies simulation or prediction while analysis refers to the larger process of identifying a question and using the results of a model to answer it. 3. A data representation of reality (for example, vector data model, TIN data model, raster data model).

M

monochromatic 1. [REMOTE SENSING] Related to a single wavelength or a very narrow band of wavelengths. 2. [GRAPHICS] One color on a contrasting background.

monument *See* survey monument.

morphology 1. The structure of a surface. 2. The study of structure or form.

morphometric A map of surface features on the earth.

mosaic 1. Maps of adjacent areas with the same projection, datum, ellipsoid, and scale whose boundaries have been matched and dissolved. *See also* edgematching. 2. An image made by assembling individual images or photographs of adjacent areas.

multiband photography *See* multispectral photography.

multichannel receiver [GPS] A receiver that tracks several satellites at a time, using one channel for each satellite. *Compare* multiplexing channel receiver.

multipart feature A geographic feature composed of more than one physical part that is stored as one object. In a layer of states, for example, Hawaii could be considered a multipart feature because it is made of many polygons but only references one set of attributes in the database.

multipath [GPS] Also **multipath error** Errors caused when a satellite signal reaches the receiver from two or more paths, one directly from the satellite and the others reflected from nearby buildings or other surfaces. Signals from satellites low on the horizon will produce more error; many receivers can mask satellite signals coming from too low an angle.

multiplexing channel receiver [GPS] A receiver that tracks several satellite signals within a single channel. *Compare* multichannel receiver.

multipoint feature A feature consisting of more than one point that only references one set of attributes in the database. *See* multipart feature.

multispectral [REMOTE SENSING] Related to two or more frequencies or wavelengths in the electromagnetic spectrum.

multispectral photography [REMOTE SENSING, PHOTOGRAMMETRY] Also **multiband photography, multispectral imaging** Photography that creates imagery from several narrow spectral bands within the visible light region and the near infrared region. A multispectral image contains two or more images, each taken from a different portion of the spectrum (e.g., blue, green, red, infrared).

multispectral scanner (MSS) [REMOTE SENSING, PHOTOGRAMMETRY] A device carried on satellites and aircraft that records radiation from the ultraviolet, visible, and infrared portions of the electromagnetic spectrum.

m-value Measure value. *See* measure.

nadir 1. [ASTRONOMY] The point on the celestial sphere directly beneath an observer. Both the nadir and the zenith lie on the observer's meridian; the nadir lies 180 degrees from the zenith and is therefore unobservable. 2. [REMOTE SENSING] In aerial photography, the point on the ground vertically beneath the perspective center of the camera lens.

National Geodetic Vertical Datum of 1929 Formerly **Mean Sea Level 1929** The datum established in 1929 by the U.S. Coast and Geodetic Survey as the surface against which elevation data in the United States is referenced.

natural breaks classification A data classification method that uses a statistical formula called Jenk's optimization to find patterns in the data by minimizing the sum of the variance within each class.

Navigation Satellite Timing and Ranging (NAVSTAR) [GPS] The full name of the Global Positioning System.

nearest neighbor assignment [DIGITAL IMAGE PROCESSING, ARC GRID] A technique for resampling raster data in which the value of each cell in an output grid is calculated using the value of the nearest cell in an input grid. Nearest neighbor assignment does not change any of the values of cells from the input layer; for this reason it is often used to resample categorical or integer data (for example, land use, soil, or forest type). *See also* bilinear interpolation, cubic convolution.

neatline [CARTOGRAPHY] A border drawn around a map to enclose the legend, scale, title, geographic features, and any other information pertinent to the map, often showing tick marks that indicate intervals of distance. On a standard quadrangle map, the neatlines are the meridians and parallels delimiting the quadrangle.

neighborhood functions Defining new values for locations using the values of other locations within a given distance or direction. *See also* proximity analysis, nearest neighbor assignment.

network 1. An interconnected set of lines representing geographic features such as roads, wires, pipes, or cables through which resources can be moved. 2. [COMPUTING] A group of computers that share software, data, and peripheral devices. *See* LAN and WAN.

network

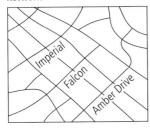

network analysis Any method of calculating locations and relationships in a network, usually in order to study or model connectivity, rate of flow, or capacity.

network nodes The connecting points in a network, for example, intersections and interchanges of a road network, confluence of streams in a hydrologic network, or switches in a power grid.

network trace A function that follows connectivity in a geometric network. Specific kinds of network tracing include finding features that are connected, finding loops, tracing upstream, and tracing downstream.

nodata In a grid or other raster format, the absence of a recorded value. While the measure of a particular attribute in a cell may be zero, a nodata value indicates that no measurements have been taken for that cell at all. *See also* null value.

node 1. The beginning and ending points of an arc, topologically linked to all the arcs that meet there. *See* network nodes, from-node, to-node. 2. In graph theory, the location at which three or more lines connect. 3. One of the three corner points of a triangle in a TIN, topologically linked to all triangles that meet there. Each sample point in a TIN becomes a node in the triangulation. 4. [COMPUTING] The point at which a computer, or other addressable device, attaches to a communications network.

nodes

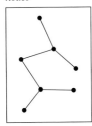

69

node attribute table (NAT) A table containing attributes for the nodes in a coverage that represent point features. The NAT contains one record for each node. At minimum, the NAT stores the internal number of each node, the feature ID of each node, and the internal number of one of the arcs to which the node is attached. *See also* feature attribute table.

node snap *See* snapping.

noise 1. [REMOTE SENSING] Any disturbance in a frequency band. 2. Any irregular, sporadic, or random oscillation in a transmission signal. 3. Random or repetitive events that interfere with communication.

nominal 1. Relating to a name. 2. Data divided into classes where no class comes before another in sequence or importance, for example, a group of polygons colored to represent different soil types. *See also* qualitative.

normal distribution Also **Gaussian distribution** The symmetrical distribution of values about a mean with a given variance, characterized by a bell curve.

normalization Dividing one numeric attribute value by another in order to minimize differences in values based on the size of areas or the number of features in each area. For example, normalizing (dividing) total population by total area yields population per unit area, or density.

North American Datum of 1927 (NAD 1927, NAD27) The primary local geodetic datum used to map the United States during the middle part of the 20th century, referenced to the Clarke spheroid of 1866 and an initial point at Meades Ranch, Kansas. Features on USGS topographic maps, including the corners of 7.5-minute quadrangle maps, are referenced to NAD27. It is gradually being replaced by the North American Datum of 1983.

North American Datum of 1983 (NAD 1983, NAD83) A geocentric datum based on the Geodetic Reference System 1980 ellipsoid (GRS80). Its measurements are obtained from both terrestrial and satellite data.

north arrow A map symbol that points north, thereby showing how the map is oriented.

north arrow

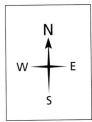

northing 1. The distance north that a point in a coordinate system lies from the origin, measured in that system's units. 2. The y-value in a rectangular coordinate system.

null value The absence of a recorded value for a geographic feature. A null value differs from a value of zero in that zero may represent the measure of an attribute, while a null value indicates that no measurement has been taken. *See also* nodata.

O

object class The storage format for nonspatial objects in a geodatabase. *Compare* feature class.

oblate ellipsoid A geometric solid made by rotating an ellipse about its shorter axis. The shape of the earth approximates an oblate ellipsoid with a flattening of one part in 298.257.

oblateness *See* flattening.

oblique aspect *See* oblique projection.

oblique photograph [PHOTOGRAMMETRY, REMOTE SENSING] A photograph taken with the axis of the camera held at an angle between the horizontal plane of the ground and the vertical plane perpendicular to the ground. A low oblique image shows only the surface of the earth; high oblique includes the horizon. *Compare* vertical photograph.

oblique projection 1. A conic projection whose axis does not line up with the polar axis of the globe. 2. A cylindrical projection whose lines of tangency or secancy follow neither the equator nor a meridian. 3. Also **oblique aspect** A planar projection whose point of tangency is neither on the equator nor at a pole.

off-nadir [REMOTE SENSING] Any point not directly beneath a scanner's detectors, but rather off at an angle. *See* nadir and zenith.

one-to-many A relationship between two linked or joined tables where one record in the first table corresponds to many records in the second table. *See also* join, link.

ordinal 1. Relating to a specified order or rank. 2. Data classified by comparative value. For example, lines on a map may be ranked in order of thickness to distinguish trails, dirt roads, paved roads, and highways.

ordinate [MATHEMATICS, COORDINATE GEOMETRY] In a rectangular coordinate system, the vertical distance of the y-coordinate from the horizontal or x-axis. For example, a point with the coordinates (7,3) has an ordinate of 3. The x-coordinate of a point is called the abscissa.

origin 1. The point in a coordinate system from which all other points are calculated, usually represented by the coordinates (0,0) in a planar coordinate system and (0,0,0) in a three-dimensional system. The center of a projection is not always its origin. 2. In a network, the beginning of a route or path.

orthocorrection Also **orthorectification** [PHOTOGRAMMETRY] Correcting distortion in satellite images caused by uneven terrain.

orthodrome *See* great circle.

orthogonal Right angled; intersecting or lying at right angles.

orthographic projection A planar projection, tangent to the earth at one point, that views the earth's surface from a point approaching infinity, as if from deep space.

orthomorphic projection *See* conformal projection.

orthophotograph Also **digital orthophoto** [REMOTE SENSING, PHOTOGRAMMETRY] A perspective aerial photograph from which distortions owing to camera tilt and ground relief have been removed. An orthophotograph has the same scale throughout and can be used as a map.

orthophotomap 1. Also **orthophotomosaic** A map made by assembling orthophotographs that cover adjacent areas, often with contour lines, color, and other cartographic symbols added, scaled to a standard reference system. 2. An orthophotograph.

orthophotoquad An orthophotograph or orthophotomap in standard quadrangle format with little or no cartographic treatment.

orthophotoscope A photomechanical or optical-electronic device that creates an orthophotograph by removing geometric and relief distortion from a vertical aerial photograph.

orthorectification *See* orthocorrection.

overlay Superimposing two or more maps registered to a common coordinate system, either digitally or on a transparent material, in order to show the relationships between features that occupy the same geographic space. *See also* topological overlay.

overlay

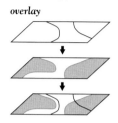

overprinting Revising a map by printing new information on top of it, usually in a distinctive color such as purple.

overshoot The portion of an arc digitized past its intersection with another arc. *See also* dangling arc.

page coordinates *See* page units.

page units Also **page coordinates** The units, usually millimeters or inches, used to arrange map elements on a page for printing, as opposed to the coordinate system on the ground that the map represents. *Compare* map units.

pan To move an on-screen display window up, down, or sideways over a map or image without changing the viewing scale.

panchromatic Sensitive to light of all wavelengths in the visible spectrum.

paneled map A map spliced together from smaller panels representing neighboring areas.

parallax [PHOTOGRAMMETRY, REMOTE SENSING, ASTRONOMY] The apparent shift in an object's position when it is viewed from two different angles.

parallel 1. Separated everywhere by the same distance. 2. [GEODESY, NAVIGATION] Also **parallel of latitude, small circle** A horizontal line encircling the earth, parallel to the equator and connecting all points of equal latitude.

parameter A variable that determines the outcome of a function or operation.

parametric curve Also **true curve** A curve that is defined mathematically rather than by a series of connected vertices. A parametric curve has only two vertices, one at each end.

parcel A tract or plot of land. The term is usually used in the context of land use or legal ownership.

parity Evenness or oddness. In address geocoding, parity is used to locate an address on the correct side of the street (for example, odd numbers on the left side, even on the right).

parse 1. [COMPUTING] To divide a sequence of letters or numbers into parts, especially to test their agreement with a set of syntax rules. 2. To break a sentence into parts of speech and describe them grammatically.

P

passive remote sensing A remote sensing system, such as aerial photography, that only detects energy naturally reflected or emitted by an object. *Compare* active remote sensing.

passive sensors [REMOTE SENSING] Imaging sensors that can only receive radiation, not transmit it.

PAT *See* point attribute table, polygon attribute table.

path 1. [NETWORK ANALYSIS] The links and nodes in a network connecting an origin to a destination. 2. [COMPUTING] Also **pathname** The location of a file, given as the drive, directories, subdirectories, and file name, in that order.

pathfinding Constructing a route between an origin and destination, most often as a least-cost path.

pathname *See* path.

Paul Revere tour A version of the traveling salesperson problem in which the starting location is different than the ending location.

P-code Also **Precise, Precision,** or **Protected code** [GPS] The pseudo-random code used by United States and allied military GPS receivers. *Compare* C/A code.

PDOP Positional Dilution of Precision. *See* Dilution of Precision.

peak A point on a surface around which all slopes are negative.

perigee [ASTRONOMY] The point in a satellite's elliptical orbit that is closest to the earth, and at which the satellite's velocity is greatest.

peripheral Any hardware device attached to a computer that the computer does not need in order to function, for example, a digitizer, plotter, printer, or scanner.

personal geodatabase A geodatabase that stores data in a single-user relational database management system (RDBMS). A personal geodatabase can be read simultaneously by several users, but only one user at a time can write data into it.

photogeology Interpreting and mapping geologic features from aerial photographs.

photogrammetry Recording, measuring, and plotting electromagnetic radiation data from aerial photographs and remote sensing systems against land features identified in ground control surveys, generally in order to produce planimetric, topographic, and contour maps.

photomap (photographic map) An aerial photograph or photographs, referenced to a ground control system and overprinted with map symbology.

photometer Also **illuminometer** An instrument that records the intensity of light by converting incident radiation into an electrical signal and then measuring it. *See also* spectrophotometer.

physical geography The study of the natural features of the earth's surface.

pie chart A chart shaped like a cut pie in which percentage values are represented as proportionally sized slices. Used to represent the relationship between parts and the whole.

pinch-roller *See* plotter.

pit A point on a surface around which all slopes are positive.

pixel (picture element) 1. [COMPUTING] The smallest addressable hardware unit on a display device. 2. The smallest unit of information in an image or raster map. Usually rectangular, pixel is often used synonymously with cell.

pixel coordinate system An image coordinate system whose measurement units are pixels. In contrast to most map coordinate systems, the origin (0,0) usually lies in the upper left corner of the image and the y-values increase as they go down the page. *Compare* Cartesian coordinate system, planar coordinate system.

planar coordinate system A two-dimensional coordinate system that locates features according to their distance from an origin (0,0) along two axes, a horizontal axis (x) representing east–west and a vertical axis (y) representing north–south.

planar enforcement *See* topology.

planar projection A projection made by projecting the globe onto a tangent or secant plane; also called azimuthal or zenithal projection as it shows true direction.

plane rectangular coordinate system *See* planar coordinate system.

plane survey A survey of a small area that does not take the curvature of the earth's surface into account.

planimetric 1. Two-dimensional; showing no relief. 2. A map that gives only the x,y locations of features and represents only horizontal distances correctly. *Compare* topographic.

P

planimetric base A two-dimensional map that serves as a guide for contour mapping, usually prepared from aerial photographs.

planimetric shift Deviations in the horizontal positions of features in an aerial photograph caused by differences in elevation. Planimetric shift causes changes in scale throughout the photograph.

plat A survey diagram, drawn to scale, of the legal boundaries and divisions of a tract of land.

plotter A device that draws an image onto paper or transparencies, either with colored pens or by drawing an image of electrostatically charged dots and fusing it onto the paper with toner. A flatbed plotter holds the paper still and draws along its x- and y-axes, a drum plotter draws along one axis and rolls the paper over a cylinder along the other axis, and a pinch roller draws along one axis and moves the paper back and forth on the other axis over small rollers.

plumb line Also **vertical line** A line that corresponds to the direction of gravity at a point on the earth's surface; the line along which an object will fall when dropped.

point Also **point feature** A single x,y coordinate that represents a geographic feature too small to be displayed as a line or area at that scale.

point

point attribute table (PAT) A table containing attributes for point coverage features. A coverage can have either a point attribute table or a polygon attribute table (also called a PAT) but not both. In addition to user-defined attributes, a PAT contains the area and perimeter of each feature (set to 0 for points), an internal sequence number, and a feature identifier. *See also* feature attribute table.

point-in-polygon [GEOPROCESSING] A spatial operation in which points from one coverage are overlaid with the polygons of another to determine which points are contained within the polygons. Each point assumes the attributes of the polygon that contains it.

point symbol *See* marker symbol.

polar aspect A planar projection centered at either the North or South Pole.

polar flattening *See* flattening.

polar orbit An orbit with an inclination of near 90 degrees that passes over each polar region.

polar radius The distance from the earth's geometric center to either pole.

polygon A two-dimensional closed figure with at least three sides that represents an area.

polygon

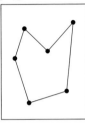

polygon–arc topology In a polygon coverage, the list of connected arcs that define the boundary of a polygon feature and the label point that links it to an attribute record in the PAT. *See* topology.

polygon attribute table (PAT) A table containing attributes for polygon coverage features. A coverage can have either a point attribute table (also called a PAT) or a polygon attribute table, but not both. In addition to user-defined attributes, a PAT contains the area and perimeter of each polygon, an internal sequence number, and a feature identifier. *See also* feature attribute table.

polygon overlay [GEOPROCESSING] Merging polygons and their attributes from two geographic data layers to make a third layer.

polyline A sequence of points, each pair of which can be connected with a straight line, a circular arc, an elliptical arc, or a Beziér curve. A polyline with a pair of points that is not connected is called a multipart polyline.

position Also **location** The latitude, longitude, and altitude (x, y, z) of a point, often accompanied by an estimate of error. It may also refer to an object's orientation (facing east, for example) without referring to its location.

Positional Dilution of Precision (PDOP) *See* Dilution of Precision.

precision 1. The number of significant digits used to store numbers, particularly coordinates. *See* single and double precision. 2. The exactness with or detail in which a value is expressed, right or wrong. *Compare* accuracy. 3. A statistical measure of repeatability, usually expressed as the variance of repeated measures about the mean.

primary colors The colors from which all other colors are derived. On a display monitor, these colors are red, green, and blue. On a color printer they are cyan, magenta, and yellow. In a painting they are red, blue, and yellow.

primary key The attribute column that uniquely identifies each row in a table, such as the unique number assigned to each parcel within a county.

prime meridian 1. The Greenwich meridian. 2. Any line of longitude designated as 0 degrees east and west, to which all other meridians are referenced.

prime meridian

prime vertical Also **prime vertical circle** [ASTRONOMY, GEODESY] The vertical circle that passes through an observer's zenith and through the east and west points of the horizon.

PRJ file *See* projection file.

project 1. *v* To display a three-dimensional surface, such as the earth, in two dimensions. *See* projection. 2. *n* In ArcView, a file that organizes the views, tables, charts, layouts, and scripts used for geographic analysis and mapmaking.

projected coordinates Latitude and longitude coordinates projected to x,y coordinates in a planar coordinate system. *Compare* geographic coordinates.

projection *See* map projection.

projection file 1. A text file containing input and output projection parameters that can be used to convert a geographic data set from one coordinate system to another. 2. Also **PRJ file** A coverage or ARC GRID file that stores the parameters for the map projection and coordinate system of a geographic data set.

projection transformation Also **projection conversion, projection change** The mathematical conversion of a map from one projected coordinate system to another, generally used to integrate maps from two or more projected coordinate systems into a GIS.

proximity analysis Also **proximity query, proximity search** 1. Selecting geographic features (points, lines, or polygons) based on their distance from other features. 2. Selecting pixels or raster cells based on their distance from other pixels or raster cells.

pseudo node A node where only two arcs connect, or where an arc connects with itself. *See* node.

pseudo nodes

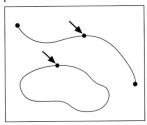

Pseudo-Random Noise (PRN) code A repeating radio signal broadcast by each GPS satellite and generated by each GPS receiver. In a given cycle, the satellite and the receiver start generating their codes at the same moment, and the receiver measures how much later the satellite's broadcast reaches it. By multiplying that time by the speed of radio waves, the receiver can compute the distance between the satellite's antenna and its own.

puck The hand-held device used with a digitizing tablet to record positions from the tablet's surface.

push broom scanner Also **along-track scanner** [REMOTE SENSING] A scanner with a line of many fixed sensors that records reflected radiation from the terrain along the satellite's direction of movement. *Compare* whisk broom scanner.

pyramid image An image format that contains successively coarser copies of an original image. The coarsest level of resolution is used to draw the entire data set. As the display zooms in, finer layers of resolution are used. Drawing speed is maintained because fewer pixels are needed to represent overviews of large areas.

quadrangle (quad) Also **topographic map, topo** A rectangular map bounded by lines of latitude and longitude, often a map sheet in either the 7.5-minute or 15-minute series published by the U.S. Geological Survey.

quadrant 1. One quarter of a circle, having an arc of 90 degrees. 2. A map having the dimensions of 15 minutes longitude by 15 minutes latitude.

quadrillage *See* grid.

qualitative 1. Data grouped by kind, not by amount or rank, such as soil by type or animals by species. 2. A map that shows only how data is distributed spatially. A dot map of all cities in the United States with no regard to size or population would be a qualitative map.

quantile classification A classification in which each group contains the same number of members.

quantitative 1. Data that can be measured, such as air temperature or wheat production. 2. A map showing the spatial distribution of measurable data, such as a map of counties shaded by population.

query Also **attribute query** A statement or logical expression used to select features or records from a database. *See* spatial query, Structured Query Language.

radar (radio detecting and ranging) A device or system that detects surface features on the earth by bouncing radio waves off them and measuring the energy reflected back.

radar altimeter An instrument that determines elevation, usually from mean sea level, by measuring the amount of time an electromagnetic pulse takes to travel from an aircraft to the ground and back again.

radian The angle subtended by an arc of a circle that is the same length as the radius of the circle, approximately 57 degrees, 17 minutes, and 44.6 seconds. A circle is 2π radians.

radiation The emission and propagation of energy through space in the form of waves. Electromagnetic energy and sound are examples of radiation.

radiometer [REMOTE SENSING] An instrument, such as an infrared radiometer or a microwave radiometer, that measures the intensity of radiation in a particular band of wavelengths in the electromagnetic spectrum.

radiometric resolution Also **radiometric sensitivity** [REMOTE SENSING] The number of digital levels that data collected by a sensor is divided into, usually expressed as a number of bits; 1-bit is two levels, 2-bit is four levels, 8-bit is 256 levels. In general, the more levels, the greater the detail. The number of levels is also called the digital number or DN value.

raster 1. A spatial data model made of rows and columns of cells. Each cell contains an attribute value and location coordinates; the coordinates are contained in the ordering of the matrix, unlike a vector structure which stores coordinates explicitly. Groups of cells that share the same value represent geographic features. *See also* grid; *compare* vector. 2. The illumination on a video display produced by repeatedly sweeping a beam of electrons over the phosphorescent screen line by line from top to bottom. 3. Also **raster image, bitmap image, image** A matrix of pixels whose values represent the level of energy reflected or emitted by the surface being photographed, scanned, or otherwise sensed.

raster

R

rasterization Also **vector-to-raster conversion** The conversion of points, lines, and polygons into cell data.

raster-to-vector conversion Also **vectorization** The conversion of cell data into points, lines, and polygons.

ratioing [REMOTE SENSING, DIGITAL IMAGE PROCESSING] Enhancing the contrast between features in an image by dividing the digital number (DN) values of pixels in one image by the corresponding DN values of pixels in a second image.

ray tracing A technique that traces imaginary rays of light from a viewer's eye to the objects in a three-dimensional scene, in order to determine which parts of the scene should be displayed from that perspective.

real-time differential GPS Differential correction performed and transmitted from a base station to a roving receiver while it is out collecting data. Differential correction performed at a later time is often referred to as "postprocessing."

record 1. A row in a database or in an attribute table that contains all of the attribute values for a single entity. 2. [COMPUTING] Also **line** An ordered set of fields in a file.

rectification [GEOREFERENCING] 1. Referencing features in an image or grid to a geographic coordinate system. 2. Converting an image or map from one coordinate system to another. 3. Removing the effects of tilt or relief from a map or image.

rectilinear 1. Straight lines, usually taken as lines that are parallel to orthogonal axes. 2. A map or image whose horizontal and vertical scales are identical.

reference datum Any datum, plane, or surface from which other quantities are measured.

reference ellipsoid An ellipsoid associated with a geodetic reference system or geodetic datum. *See* ellipsoid, geoid.

reference spheroid *See* reference ellipsoid.

region A coverage feature class that can represent a single area feature as more than one polygon.

register 1. To align two or more maps or images so that equivalent geographic coordinates coincide. 2. To link map coordinates to ground control points.

relate Temporarily connecting records in two tables using an item common to both. *Compare* relational join.

relate key The set of columns, or items, used to relate two attribute tables. *See also* primary key and foreign key.

relation *See* table.

relational database Data stored in tables that are associated by shared attributes. Any data element can be found in the database through the name of the table, the attribute (column) name, and the value of the primary key. In contrast to hierarchical and network database structures, the data can be arranged in different combinations.

relational database management system (RDBMS) *See* relational database.

relational join Permanently merging two attribute tables using an item common to both. *Compare* relate.

relational operators Phrases used to compare values associated with spatial data: greater than, less than, maximum, minimum, contains, and so forth.

relative accuracy 1. Accuracy with respect to a known point or points. 2. Of a map, its accuracy in relation to a local datum. *Compare* absolute accuracy.

relative coordinates Coordinates identifying the position of a point with respect to another point.

relief Elevations and depressions of the earth's surface, including those of the ocean floor. Relief can be represented on maps by contours, shading, hypsometric tints, digital terrain modeling, or spot elevations.

relief map A map that is or appears to be three-dimensional.

relief shading *See* hillshading.

remote sensing Collecting and interpreting information about the environment and the surface of the earth from a distance, primarily by sensing radiation that is naturally emitted or reflected by the earth's surface or from the atmosphere, or by sensing signals transmitted from a satellite and reflected back to it. Examples of remote sensing methods include aerial photography, radar, and satellite imaging.

R

remote sensing imagery Imagery acquired from satellites and aircraft. Examples include panchromatic, infrared black-and-white, and infrared color photographs, and thermal infrared, radar, and microwave imagery.

representative fraction (RF) The ratio of a distance on a map to the equivalent distance measured in the same units on the ground. A scale of 1:50,000 means that one inch on the map equals 50,000 inches on the ground. *See also* scale.

resampling 1. Reducing the file size of an image or a grid by representing a group of pixels with a single pixel. A resampled image appears coarser than the image it is taken from because it uses less information to represent the same geographic extent. 2. Transforming a raster image to a particular scale and projection.

resolution 1. The area represented by each pixel in an image. 2. The smallest spacing between two display elements, expressed as dots per inch, pixels per line, or lines per millimeter. 3. The detail with which a map depicts the location and shape of geographic features. The larger the map scale, the higher the possible resolution. As scale decreases, resolution diminishes and feature boundaries must be smoothed, simplified, or not shown at all; for example, small areas may have to be represented as points.

resolution merging Sharpening a low-resolution multiband image by merging it with a high-resolution monochrome image.

RGB Red, green, and blue, the primary additive colors used to display images on a monitor. RGB colors are produced by emitting light, rather than by absorbing it as is the case with ink on paper. Adding 100 percent of these colors results in white. *See also* CMYK.

rhumb line Also **loxodrome, loxodromic curve** A line that shows true direction on the earth's surface, crossing all meridians at the same angle.

ring The boundary, but not the space within it, represented by a line or a set of joined lines that closes on itself.

roamer [NAVIGATION, SURVEYING] A transparent gauge that represents easting and northing distances at a given map scale, used to locate positions on a map.

root mean square (RMS) error A measure of the difference between locations that are known and locations that have been interpolated or digitized. The RMS error is derived by squaring the differences between known and unknown points, adding those together, dividing that by the number of test points, and then taking the square root of that result.

route A path through a network or grid from a source to a destination.

route

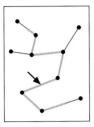

route attribute table (RAT) A table that stores a sequence number, feature identifier, and any user-defined attributes for each route in a route system. There is one RAT for each route system in a coverage. *See also* feature attribute table.

route event An event that is associated with a route system. There are three types of route events: linear, continuous, and point. An example of a linear event would be a closed left lane on route I-75 from the 31.5 to the 32.1 mileposts. An example of a continuous event would be the speed limits along a highway, where the start position of one event is the same as the end position of the preceding event. An example of a point event would be an accident at milepost 6.3 on route I-64.

route system A collection of routes. A single linear data set can contain several route systems; for example, a road layer can contain a bus route system, a highway route system, and a pizza delivery route system. *See also* route event, dynamic segmentation.

routing analysis *See* network analysis.

rover Also **mobile receiver** A portable GPS receiver used to collect data in the field. The rover's position can be computed relative to a second, stationary GPS receiver.

row 1. A horizontal record in an attribute table. 2. A horizontal group of cells in a grid, or pixels in an image.

rubber banding *See* rubber sheeting.

rubber sheeting Also **warping, elastic transformation** Mathematically stretching or shrinking a portion of a map or image in order to align its coordinates with known control points.

run-length encoding A data compression technique for storing raster or grid data. Run-length encoding stores data by row. If two or more adjacent cells in a row have the same value, the database stores that value instead of recording a separate value for each cell. The more adjacent columns there are with the same value, the greater the compression.

run time [COMPUTING] 1. The time during which a program is running. 2. The time it takes to run a program.

S

satellite constellation 1. The arrangement in space of a set of satellites. 2. All the satellites visible to a GPS receiver at one time. 3. The set of satellites that a GPS receiver uses to calculate positions.

satellite imagery *See* remote sensing imagery.

saturation 1. Also **intensity, richness, chroma** How pure a color is; the perceived amount of white in a hue relative to its brightness, or how free it is of gray of the same value. 2. [REMOTE SENSING] Where energy flux exceeds the sensitivity range of a detector.

scale The ratio or relationship between a distance or area on a map and the corresponding distance or area on the ground. *See* bar scale, verbal scale, representative fraction.

scale

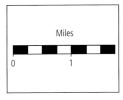

scale bar *See* bar scale.

scale factor 1. The ratio of the actual scale at a particular place on a map to the stated scale of the map. 2. A value, usually less than one, that converts a tangent projection to a secant projection.

scanner 1. A device that sweeps a light beam across the surface of a map or image and records the information in raster format. *See also* drum scanner, flatbed scanner. 2. A device that records the radiation reflected or emitted by the earth's surface and processes it as per (1).

scatter chart A chart in which each data point is marked with its own symbol against perpendicular x- and y-axes.

scratch file [COMPUTING] A file, created by either a user or an operating system, that holds temporary data or results during an operation. When the operation is complete, the file is deleted.

script A set of instructions for an application program, usually written in the application's syntax. *See also* macro.

scrubbing 1. Checking the accuracy of data before it is converted into a different format. 2. Improving the appearance of scanned or digitized data by closing open polygons, fixing overshoots and undershoots, refining thick lines, and so forth.

secant A straight line that cuts a curve or surface at two or more points.

secant projection A projection whose surface intersects the surface of the earth. A secant conic or cylindrical projection, for example, is recessed into the globe, intersecting it at two circles. At the lines of intersection the projection is free from distortion. *Compare* tangent projection.

second 1. Also **angular second, arc-second, second of arc** An angle equal to one sixtieth of a minute of arc. 2. One sixtieth of a minute of time.

section 1. The arcs or portions of arcs used to define a route. 2. One thirty-sixth of a township, bounded by parallels and meridians, equal to one square mile and containing 640 acres.

section table (SEC) The attribute table for the section feature class in a coverage, containing the route number and arc number to which the section belongs, the starting and ending positions expressed as percentages of the arc length, the starting and ending positions expressed as measures along the route, an internal sequence number, a section feature identifier, and user-defined attributes. *See also* feature attribute table.

segment A line that connects vertices.

selective availability (S/A) The intentional degradation by the U.S. Department of Defense of the GPS signal for civilian receivers, which can cause errors in position of up to 100 meters. Selective availability was removed from the civilian signal in May 2000.

semimajor axis The equatorial radius of a spheroid, often referred to as "a."

semiminor axis The polar radius of a spheroid, often referred to as "b."

sensor An electronic device for detecting energy, whether emitted or radiated, and converting it into a signal that can be recorded and displayed as numbers or as an image.

sextant A hand-held navigational instrument that measures, from its point of observation, the angle between a celestial body and the horizon or between two objects. The angle is measured on a graduated arc that covers one sixth of a circle (60 degrees).

sextant

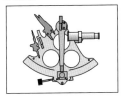

shaded relief image A raster image that shows light and shadow on terrain from a given angle of the sun.

shade symbol A color or pattern for filling polygons on a map.

shading Graphic patterns such as cross hatching, lines, or color or grayscale tones that distinguish one area from another on a map.

shape The visible form of a geographic object. Most geographic objects can be represented on a map using three basic shapes: points, lines, and polygons.

shapefile A vector file format for storing the location, shape, and attributes of geographic features. It is stored in a set of related files and contains one feature class.

sheet lines Also **neatlines** The border of a map, usually composed of parallels and meridians.

shortest path analysis Determining the route of least impedance between two points, taking into account topography and travel restrictions such as one-way streets and rush-hour traffic.

signal 1. The modulation of an electric current, electromagnetic wave, or other type of flow in order to convey information. 2. The current or wave itself. 3. The information itself.

signature *See* spectral signature.

simplification The part of cartographic generalization that eliminates the less essential details from a map whose scale has been reduced.

single precision A level of coordinate accuracy that stores up to seven significant digits for each coordinate, retaining a precision of 5 meters in an extent of 1,000,000 meters. *See also* double precision.

sink *See* depression contour.

sliver polygons Small, narrow polygon features that inevitably appear along the borders of polygons following the overlay of two or more geographic data sets.

sliver polygon

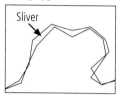

slope 1. An inclined surface. A slope may be concave, straight, convex, or any combination thereof. 2. A measure of change in surface value over distance, expressed in degrees or as a percentage. For example, a rise of 2 meters over a distance of 100 meters describes a 2-percent slope.

slope image A raster image showing change in elevation, usually color-coded to indicate how steep the slope is at each pixel.

small circle The circle made when a flat plane intersects a sphere anywhere but through the center of the sphere. Parallels of latitude other than the equator are small circles.

small scale Generally, a map scale whose representative fraction is 1:500,000 or smaller. A small-scale map shows a relatively large area on the ground with a low level of detail. *See* scale, representative fraction.

smoothing [DIGITAL IMAGE PROCESSING] Reducing or removing small variations in an image to reveal the global pattern or trend, either through interpolation or by passing a filter over the image. *See also* low-pass filter.

snapping Moving a feature, or a portion of it, to coincide with the coordinates of another feature. *See also* undershoot, overshoot.

snapping distance The distance within which snapping occurs between points or lines, based on data location.

soundex A phonetic spelling (up to six characters) of a street name, used for address matching. Each of the twenty-six letters in the English alphabet is replaced with a letter in the soundex equivalent:

English: A B C D E F G H I J K L M N O P Q R S T U V W X Y Z
soundex: A B C D A B C H A C C L M M A B C R C D A B W C A C

Where possible, geocoding uses a soundex equivalent of street names for faster processing. Candidate street names are initially found using soundex, then their real names are compared and verified.

space coordinate system Also **space rectangular coordinates** A three-dimensional, rectangular, Cartesian coordinate system that has not been adjusted for the earth's curvature. The x- and y-axes lie in a plane tangent to the earth's surface and the z-axis points upward.

spaghetti digitizing Digitizing that does not identify intersections as it records lines. Spaghetti digitizing is typically used to define straight lines. For precise features, or those that curve and twist, discrete digitizing is preferred.

spatial analysis Studying the locations and shapes of geographic features and the relationships between them. It traditionally includes overlay and contiguity analysis, surface analysis, linear analysis, and raster analysis.

spatial data 1. Information about the locations and shapes of geographic features, and the relationships between them; usually stored as coordinates and topology. 2. Any data that can be mapped.

spatial feature *See* geographic feature.

spatial join [GEOPROCESSING] Joining the attributes of features in two different geographic layers based on the relative locations of the features.

spatial modeling Any procedures that use the spatial relationships between geographic features to simulate real-world conditions, such as geometric modeling (generating buffers, calculating areas and perimeters, and calculating distances between features), coincidence modeling (topological overlay), and adjacency modeling (pathfinding, redistricting, and allocation).

spatial query Selecting geographic features by where they are in relation to each other. For example, features can be selected if they fall inside, are adjacent to, or lie within a specified distance of other features.

spatial reference system A system with a point of origin, units of measure, and reference axes for locating positions on the earth.

S

spatial resolution *See* resolution.

spectral resolution Also **bandwidth** The range of wavelengths that a satellite imaging system can detect.

spectral signature The pattern of electromagnetic radiation (spectral lines) that identifies a chemical or compound.

spectrometer *See* spectrophotometer.

spectrophotometer Also **spectroradiometer, spectroradiophotometer, spectrometer** A photometer that measures the intensity of electromagnetic radiation as a function of its frequency. It is usually used for measuring the visible portion of the spectrum.

spectroscopy The study of how electromagnetic radiation is absorbed and reflected.

spectrum *See* electromagnetic spectrum.

sphere A three-dimensional shape whose center is equidistant from every point on its surface, made by revolving a circle around its diameter.

spherical coordinate system A system of latitude and longitude that defines the locations of points on the surface of a sphere or spheroid. Distances east–west are measured with lines that run north and south (longitude or meridians) and distances north–south are measured with lines that run east and west (latitude or parallels).

spheroid Also **rotational ellipsoid, ellipsoid of revolution** 1. A three-dimensional shape obtained by rotating an ellipse about its minor axis, resulting in an oblate spheroid, or about its major axis, resulting in a prolate spheroid. 2. When used to represent the earth, a spheroid as defined in (1), but with dimensions that either approximate the earth as a whole, or with a part that approximates the corresponding portion of the geoid.

spike 1. An overshoot line created erroneously by a scanner and its rasterizing software. 2. An anomalous data point that protrudes above or below an interpolated surface.

spline 1. A mathematical curve that is used to smoothly represent variation, either in a line or on a surface. 2. To create a curve in a line by inserting vertices. *Compare* densify.

spline

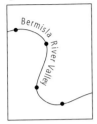

spot elevation Also **spot height** A small dot or symbol on a map marking a surveyed elevation, usually according to its height above a particular datum. *See* elevation, contour line.

spurious polygon *See* sliver polygon.

SQL *See* Structured Query Language.

stable base In cartography, any material such as Mylar® or film that is more durable than paper and less likely to shrink or stretch.

standard deviation [STATISTICS] A measure of the spread of values from their mean, calculated as the square root of the arithmetic mean of the squares of the deviations from the mean.

standard line A line on a sphere or spheroid that stays the same length after being projected, commonly a standard parallel or central meridian.

standard parallel The line of latitude in a conic or cylindrical projection where the cone or cylinder touches the globe. A tangent conic or cylindrical projection has one standard parallel, while a secant conic or cylindrical projection has two. At the standard parallel, the projection shows no distortion.

State Plane coordinate system (SPCS) A group of planar coordinate systems that divides the United States into more than 130 zones, so that distortion in each is less than one part in 10,000. Each zone has its own map projection and parameters and uses either the NAD27 or NAD83 horizontal datum. The Lambert conformal conic projection is used for states that extend mostly east–west, while transverse Mercator is used for those that extend mostly north–south. The oblique Mercator projection is used for the panhandle of Alaska.

static positioning [GPS] Determining a position on the earth by averaging the readings taken by a stationary antenna over a period of time. *Compare* kinematic positioning.

S

statistical surface Ordinal, interval, or ratio data represented as a surface. The height of each area is proportional to a numerical value.

steradian The solid angle subtended at the center of a sphere of radius r by a bounded region on the surface of the sphere having an area r squared. There are 4π steradians in a sphere.

stereocompilation [PHOTOGRAMMETRY] A map produced with a stereoscopic plotter using aerial photographs and geodetic control data.

stereographic projection 1. A tangent planar projection that views the earth's surface from a point on the globe opposite the tangent point. 2. A secant planar projection that views the earth from a point on the globe opposite the center of the projection.

stereometer Also **parallax bar** A stereoscope containing a micrometer for measuring the effects of parallax in a stereoscopic image.

stereomodel Also **stereoscopic model** [PHOTOGRAMMETRY] The three-dimensional image formed where rays from points in the images of a stereoscopic pair intersect.

stereopair Also **stereoscopic pair, aerial stereopair** [PHOTOGRAMMETRY] Two aerial photographs of the same area taken from slightly different angles that when viewed together through a stereoscope produce a three-dimensional image.

stereoplotter Also **stereoscopic plotter** An instrument that projects a stereoscopic image from aerial photographs, converts the locations of objects and landforms on the image to x,y,z coordinates, and plots these coordinates as a drawing or map.

stereoscope A binocular device that produces the impression of a three-dimensional image from two overlapping images of the same area.

stereoscopic pair *See* stereopair.

stop impedance [NETWORK ANALYSIS] The time it takes for a stop to occur, used to compute the impedance of a path or tour.

string 1. A series of letters or numbers, or both, enclosed by quotes, sometimes with a fixed length. 2. A set of coordinates that defines a group of linked line segments. *See* line and arc.

Structured Query Language (SQL) A syntax for defining and manipulating data in a relational database. Developed by IBM in the 1970s, it has become an industry standard for query languages in most relational database management systems.

subtractive primary colors The three primary colors, cyan, magenta, and yellow, that when used as filters for white light remove blue, green, and red light, respectively.

surface A geographic phenomenon represented as a set of continuous data, such as elevation or air temperature. Models of surfaces can be built from sample points, isolines, bathymetry, and the like. *See also* surface model.

surface fitting Generating a statistical surface that approximates the values of a set of known x,y,z points.

surface model A digital abstraction or approximation of a surface, generalized from sample data and housed in a data structure such as a grid, lattice, or TIN.

surveying Measuring physical, chemical, or geometric characteristics of the earth. Surveys are often classified by the type of data studied or by the instruments or methods used. Examples include geodetic, geologic, topographic, hydrographic, land, geophysical, soil, mine, and engineering surveys.

survey monument Also **survey marker** An object placed at the site of a survey station.

survey station A location on the earth that has been accurately determined by geodetic survey.

symbol [CARTOGRAPHY] A mark used to represent a geographic feature on a map. Symbols can look like what they represent (tiny trees, railroads, houses) or they can be abstract shapes (points, lines, polygons). They are usually explained in a map legend.

symbolization Devising a set of marks of appropriate size, color, shape, and pattern, and assigning them to map features to convey their characteristics or their relationships to each other at a given map scale.

syntax The structural rules for using statements in a command or programming language.

table *Also* **relation** Data arranged horizontally in rows and vertically in columns in a relational database system. A table has a specified number of columns but can have any number of rows. Rows stored in a table are structurally equivalent to records from flat files in that they must not contain repeating fields.

tablet *See* digitizer.

tabular data Descriptive information that is stored in rows and columns and can be linked to map features. *See also* spatial data.

tag 1. A label, code, or classification symbol for identifying each feature in a geographic layer. 2. [PROGRAMMING] Characters that contain information about a file or record type.

tangent projection A projection whose surface touches the earth's without piercing it. A tangent planar projection touches the globe at one point, while tangent conic and cylindrical projections touch the globe along a line. At the point or line of tangency the projection is free from distortion. *Compare* secant projection.

TDOP Time Dilution of Precision. *See* Dilution of Precision.

tesselation Dividing a two-dimensional area into polygonal tiles, or a three-dimensional area into polyhedral blocks, in such a way that no figures overlap and there are no gaps. *See* Thiessen polygons.

text attribute table (TAT) A table containing text attributes, such as color, font, size, location, and placement angle, for an annotation subclass in a coverage. In addition to user-defined attributes, the TAT contains a sequence number and text feature identifier. *See also* feature attribute table.

text envelope A rectangle that bounds a text string.

text label *See* label.

text symbol A text style defined by font, size, character spacing, color, and so on, used to label maps and geographic features.

thematic data Features of one type that are generally placed together in a single geographical layer. *See* theme.

theme 1. A vector layer of related geographic features, such as streets, rivers, or parcels, that when juxtaposed with other themes can be used in overlay analysis. 2. A raster layer of geographic information, such as an image or a grid.

theme-on-theme selection An operation that selects features in one theme using the features in another in order to answer questions about the spatial relationships between them, such as whether one feature lies within another, whether it completely contains another, or whether it is within a specified distance of another.

theodolite A surveying instrument for measuring vertical and horizontal angles, consisting of an alidade, a telescope, and graduated circles mounted vertically and horizontally.

Thiessen polygons Also **Voronoi diagrams, Dirichlet tessellations** Polygons generated from a set of points, defined by the perpendicular bisectors of the lines between all points and drawn so that each polygon bounds the region that is closer to one point than to any adjacent point.

Thiessen polygons

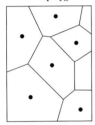

thinning *See* weeding.

three-dimensional shape A point, line, or polygon that stores x-, y-, and z-coordinates as part of its geometry. A point has one set of z-coordinates; lines and polygons have z-coordinates for each vertex.

tick marks Short, regularly spaced lines along the edge of an image or neatline that indicate intervals of distance.

tics 1. Also **ground control points** Points on a map representing locations whose coordinates are known in some system of ground measurement such as latitude and longitude. 2. Points in a data layer representing known locations, used to register map sheets for digitizing and to transform digitizer coordinates to a common coordinate system.

T

tidal datum A local datum defined by a particular phase of the tide from which heights or depths are reckoned.

tie point 1. A point whose location is determined by a tie survey. 2. A point common to the area where two or more adjacent strips of photographs overlap, used to link the images and establish a common scale between them.

tie survey A survey that uses a point of known location on the ground to determine the location of a second point.

TIGER (Topologically Integrated Geographic Encoding and Referencing) The nationwide digital database developed for the 1990 census, succeeding the DIME format. TIGER files contain street address ranges, census tracts, and block boundaries.

tile 1. A division of data within a map library, referenced by location. A tile can either be a regular shape, such as a map sheet, or irregular, such as a county border. Splitting a geographical area into tiles makes information easier to retrieve. 2. A cell in a grid, usually accompanied by attribute information.

TIN *See* triangulated irregular network.

to-node Of an arc's two endpoints, the last one digitized. *See also* from-node. From- and to-nodes give an arc left and right sides, and therefore direction. *See* topology.

topographic 1. Having elevation. 2. A map showing relief, often as contour lines, along with other natural and human-made features. 3. Map sheets published by the U.S. Geological Survey in the 7.5-minute or 15-minute quadrangle series.

topography The shape or configuration of the land, represented on a map by contour lines, hypsometric tints, and relief shading.

topological overlay Superimposing two or more geographic data sets in order to produce a new geographic layer with a new set of attributes. The geometry and the attributes of the output data layer depend on the type of overlay used. *See* Identity, Intersect, Union.

topology 1. The spatial relationships between connecting or adjacent features in a geographic data layer. Topological relationships are used for spatial modeling operations that do not require coordinate information. *See* arc–node topology, polygon–arc topology. 2. [GEOMETRY, MATHEMATICS] The branch of geometry that deals with the properties of a figure that remain unchanged even when the figure is bent, stretched, or otherwise distorted.

topology

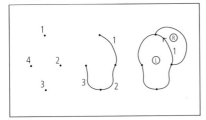

toponym A place name.

tour *See* Hamiltonian circuit.

township 1. A quadrangle approximately 6 miles on a side, bounded by meridians and parallels and containing thirty-six sections. 2. A governmental subdivision, which may vary from the standard size and shape.

transformation 1. Also **rectification** Converting the coordinates of a map or an image from one system to another, typically by shifting, rotating, scaling, skewing, or projecting them. 2. Converting data from one format to another.

translation 1. Adding a constant value to a coordinate. *See also* transformation. 2. Converting data from one format to another, usually in order to move it from one system to another.

transverse aspect A map projection whose line of tangency is oriented along a meridian rather than along the equator.

traveling salesperson problem (TSP) A tour problem in which a salesperson must find the most efficient way to visit a series of stops, then return to the starting location. In the original version of the problem, each stop may be visited only once.

tree *See* hierarchical database.

T

triangulated irregular network (TIN) A vector data structure that partitions geographic space into contiguous, nonoverlapping triangles. The vertices of each triangle are data points with x-, y-, and z-values; elevation values at these points are interpolated to create a continuous surface.

triangulated irregular network

triangulation [SURVEYING, NAVIGATION] Locating positions on the earth's surface using the principle that if the measures of one side and the two adjacent angles of a triangle are known, the other dimensions of the triangle can be determined. Surveyors begin with a known length, or baseline, and from each end use a theodolite to measure the angle to a distant point, forming a triangle. Once the lengths of the two sides and the other angle are known, a network of triangles can be extended from the first. *Compare* trilateration.

trilateration [SURVEYING, NAVIGATION] Determining the position of a point on the earth's surface with respect to two other points by measuring the distances between all three points. *Compare* triangulation.

true-direction projection Also **azimuthal projection** A projection that gives the directions, or azimuths, of all points on the map correctly with respect to the center. True-direction projections may also be conformal, equal-area, or equidistant.

true north Also **geographic north** The direction from any point on the earth's surface to the north geographic pole. *Compare* magnetic north.

tuple Also **record** A row in a relational table or database.

turn impedance The cost of making a turn at a network node. The impedance for making a left turn, for example, can be different from the impedance for making a right turn or a U-turn at the same place.

turntable A table that stores the cost of making each turn in a network. It identifies the node where the turn takes place, the line that it comes from, and the line that it turns onto.

𝒰

undershoot A line that falls short of another line that it should intersect. *See also* dangling arc.

undevelopable surface A surface, such as the earth's, that cannot be flattened into a map without stretching, tearing, or squeezing it. To produce a flat map of the round earth, its three-dimensional surface must be projected onto a developable shape such as a plane, cone, or cylinder. *See* projection.

Union [GEOPROCESSING] An overlay of two polygon coverages that preserves the features and attributes of each. *See also* Intersect, Identity.

Union

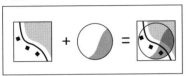

United States Geological Survey (USGS) The national mapping agency of the United States that produces paper maps, digital maps, and DEMs at a variety of scales, including 1:24,000, 1:100,000, 1:250,000, and 1:1 million. Its national map database consists of 1:100,000 maps, available as digital line graph (DLG) and TIGER files.

universal polar stereographic (UPS) A projected coordinate system that covers all regions not included in the UTM coordinate system; that is, regions above 84 degrees north and below 80 degrees south. Its central point is either the North or South Pole. *See also* universal transverse Mercator.

Universal time (UT) *See* Greenwich mean time.

universal transverse Mercator (UTM) A commonly used projected coordinate system that divides the globe into sixty zones, starting at −180 degrees longitude. Each zone extends north–south from 84 degrees north to 80 degrees south, spans 6 degrees of longitude, and has its own central meridian. *See also* universal polar stereographic.

UTM *zones*

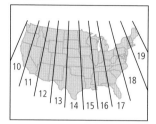

universe polygon Also **external polygon** The first record in a polygon attribute table, representing the area beyond the outer boundary of the coverage.

unprojected coordinates *See* geographic coordinates.

UTC Coordinated Universal Time. *See* Greenwich mean time.

valency 1. The number of arcs that begin or end at a node. 2. [CHEMISTRY] A measure of an element's ability to combine with other elements, expressed as the number of hydrogen atoms it can unite with or displace.

valency table A table that lists the nodes in a data layer along with their valencies.

value 1. The lightness or darkness of a color. 2. The brightness of a color or how much light it reflects; for instance, blue, light blue, dark blue.

value attribute table (VAT) A table containing attributes for a grid, including user-defined attributes, the values assigned to cells in the grid, and a count of the cells with those values.

variance The total amount of disagreement between numbers, calculated as the mean of the squares of the deviations from the mean value of a range of data.

VDOP Vertical Dilution of Precision. *See* Dilution of Precision.

vector 1. A data structure used to represent linear geographic features. Features are made of ordered lists of x,y coordinates and represented by points, lines, or polygons; points connect to become lines, and lines connect to become polygons. Attributes are associated with each feature (as opposed to a raster data structure, which associates attributes with grid cells). *Compare* raster. 2. Any quantity that has both magnitude and direction.

vector

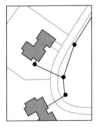

vectorization The conversion of cell or raster data into points, lines, and polygons.

vector-to-raster conversion *See* rasterization.

verbal scale A map scale that expresses the relationship between distance on the map and distance on the ground in words; for example, "One inch represents 20 miles." The units on the map do not have to be the same as the units on the ground. *Compare* bar scale, representative fraction.

vertex 1. One of a set of ordered x,y coordinates that constitutes a line. 2. The junction of lines that form an angle. 3. The highest point of a feature.

vertex

vertical control [GEODESY, SURVEYING] Control points on the ground whose elevations have been accurately surveyed in reference to the geoid, used to provide elevations for other surveys.

vertical control datum Any level surface from which elevations are reckoned, often mean sea level.

vertical exaggeration A multiplier applied uniformly to the z-values in a three-dimensional model to enhance or minimize the natural variations of its surface. Vertical exaggeration is generally applied more to flat regions than to mountainous ones.

vertical line Also **plumb line** A line that corresponds to the direction of gravity at a point on the earth's surface; the line along which an object will fall when dropped.

vertical photograph An aerial photograph taken with the camera lens pointed straight down.

viewshed A map that shows which areas are visible and which are not from a specified x,y,z position.

virtual table A logical table that stores pointers to data, not the data itself. It also identifies the order of the fields, which ones are visible, and any aliases for field names.

visual center Also **optical center** The point on a rectangular map or image to which the eye is drawn. It lies slightly (about 5 percent of the total height) above the geometric center of the page.

visual hierarchy Also **stereogrammic organization** The presentation of features on a map so that they appear to lie at different levels, either of elevation or importance. For example, land can be made to look higher than the water next to it, and monuments can be made to stand out from their surrounding plane.

Voroni diagram *See* Thiessen polygons.

voxel A three-dimensional pixel; bulky to store.

V

WAN *See* wide area network.

warping Also **rubber sheeting** Mathematically stretching or shrinking a portion of a map or image in order to register its coordinates with known control points. *See also* edgematching.

watch file A text file that records all dialog during an ArcInfo session. Watch files can be edited and converted to macro programs.

watershed [HYDROLOGY] The area drained by a river and its tributaries.

wavelength The distance between two successive crests on a wave, calculated as the velocity of the wave divided by its frequency.

weeding Also **line thinning** Reducing the number of points that define a line while preserving its essential shape. Weeding can be applied to both vector and raster data. *Compare* line smoothing.

weed tolerance The minimum distance allowed between any two vertices along a line, set before digitizing. When new lines are added, vertices that fall within that distance of the last vertex are ignored. Nodes are always retained.

weight [STATISTICS] 1. A number that tells how important a variable is for a particular calculation. The larger the weight assigned, the more that variable will influence the outcome of the operation. 2. The number of values in a set.

weighted moving average The value of a point's attribute computed by averaging the values of its surrounding points, taking into account their importance or their distance from the point.

whisk broom scanner Also **across-track scanner** [REMOTE SENSING] A scanner with an oscillating mirror that moves back and forth across the satellite's direction of travel, creating scan line strips that are contiguous or that overlap slightly. *Compare* push broom scanner.

wide area network (WAN) A computer network that operates across public or dedicated telephone lines and connects terminals in different cities or countries. *See also* local area network.

wireframe A three-dimensional picture of an object, composed entirely of lines (wires). The lines represent the edges or surface contours, including those that would otherwise be hidden by a solid view. Wireframes make editing easier as the screen redraws much more quickly.

wizard [GUI] A series of interfaces that lead a user through the steps of a long or complicated task.

workspace An ArcInfo directory that contains coverages, grids, or TINs.

workstation A computer that has better graphics and more processing power than most personal computers, and is able to carry out several tasks at once. It usually shares data and software with other computers in a network.

world file A text file containing the coordinate and scaling information for converting the coordinates in an image (which usually start at [0,0] and extend to the image's size in pixels) to map coordinates such as latitude–longitude or State Plane. A world file allows an image to be overlaid with vector data of the same geographic area.

World Geodetic System of 1972 (WGS72) A geocentric datum and coordinate system designed by the United States Department of Defense, no longer in use. *See* World Geodetic System of 1984.

World Geodetic System of 1984 (WGS84) The most widely used geocentric datum and geographic coordinate system today, designed by the U.S. Department of Defense to replace WGS72. GPS measurements are based on WGS84.

x-axis 1. In a planar coordinate system, the horizontal line that runs to the right and left (east–west) of the origin (0,0). Numbers to the east of the origin are positive and numbers to the west are negative. 2. In a spherical coordinate system, the x-axis is in the equatorial plane and passes through 0 degrees longitude. *See* y-axis, z-axis, Cartesian coordinate system. 3. On a chart, the horizontal axis.

x,y coordinates A pair of numbers expressing a point's horizontal and vertical distance along two orthogonal axes, from the origin (0,0) where the axes cross. Usually, the x-coordinate is measured along the east–west axis and the y-coordinate is measured along the north–south axis.

x,y,z coordinates In a planar coordinate system, three coordinates that locate a point by its distance from an origin (0,0,0) where three orthogonal axes cross. Usually, the x-coordinate is measured along the east–west axis, the y-coordinate is measured along the north–south axis, and the z-coordinate measures height or elevation.

y-axis 1. In a planar coordinate system, the vertical line that runs above and below (north and south of) the origin (0,0). Numbers north of the origin are positive and numbers south of it are negative. 2. In a spherical coordinate system, the y-axis lies in the equatorial plane and passes through 90 degrees east longitude. *See* x-axis, z-axis, Cartesian coordinate system. 3. On a chart, the vertical axis.

y-coordinate *See* x,y and x,y,z coordinates.

z-axis In a spherical coordinate system, the vertical line that runs parallel to the earth's rotation, passing through 90 degrees north latitude, and perpendicular to the equatorial plane, where it crosses the x- and y-axes at the origin (0,0,0). *See* x-axis, y-axis.

z-coordinate *See* x,y,z coordinates, z-value.

zenith The point on the celestial sphere directly above an observer. *Compare* nadir.

zenithal projection Also **azimuthal projection, true-direction projection** A projection that preserves direction, made by projecting the earth onto a tangent or secant plane. *See also* planar projection.

zoom To display a larger or smaller region of an on-screen map or image. *See also* pan.

z-value Also **z-coordinate** The value for a given surface location that represents an attribute other than position. In an elevation or terrain model, the z-value represents elevation; in other kinds of surface models it represents the density or quantity of a particular attribute.

A note from the editor

To keep up with the evolving field of GIS, *The ESRI Press Dictionary of GIS Terminology* will be revised and expanded regularly. We welcome suggestions for new terms from the GIS community and from professionals in other disciplines relevant to spatial mapping and analysis. If you have suggestions for the next edition, please write to *dictionary@esri.com*.

Acronyms

AGI (Association for Geographic Information)

AIRSAR (Airborne Synthetic Aperture Radar)

AIS (Airborne Imaging Spectrometer)

ASCII (American Standard Code for Information Interchange)

AVIRIS (Airborne Visible and Infrared Imaging Spectrometer)

BLM (Bureau of Land Management)

BSI (British Standards Institute)

BURISA (British Urban and Regional Information Systems Association)

CASE (Computer-Aided Software Engineering)

CGI (Common Gateway Interface)

CGIS (Canadian Geographical Information System)

CGM (Computer Graphics Metafile)

COM (Common Object Model)

CSSM (Content Standards for Spatial Metadata)

DIGEST (Digital Geographic Information Exchange Standard)

DLL (Dynamic Link Library)

DMA (Defense Mapping Agency)

DOD or DoD (Department of Defense)

DOMSAT (Domestic Satellite)

DOQ (Digital Orthophoto Quadrangle)

DOQQ (Digital Orthophoto Quarter-Quadrangle)

DXF (Drawing Interchange Format)

EDAC (Earth Data Analysis Center)

EOS (Earth Observing System)

EPA (Environmental Protection Agency)

EPS (Encapsulated PostScript®)

EROS (Earth Resources Observation Systems)

ESIC (Earth Science Information Center)

FEMA (Federal Emergency Management Agency)

FGCC (Federal Geodetic Control Committee)

FGDC (Federal Geographic Data Committee)

FIPS (Federal Information Processing Standards)

FTP (File Transfer Protocol)

GCRP (Global Change Research Program)

GIRAS (Geographic Information Retrieval and Analysis System)

GOES (Geostationary Operational Environmental Satellite)

GRASS (Geographical Resource Analysis Support System)

HTML (HyperText Markup Language)

HTTP (HyperText Transfer Protocol)

ISO (International Standards Organization)

MOSS (Map Overlay Statistical System)

NAPP (National Aerial Photography Program)

NASA (National Aeronautics and Space Administration)

NCGIA (National Center for Geographic Information and Analysis)

NIMA (National Imagery and Mapping Agency)

NOAA (National Oceanic and Atmospheric Administration)

NSDI (National Spatial Data Infrastructure)

ODBC (Open Database Connectivity)

RAR (Real Aperture Radar)

SAR (Synthetic Aperture Radar)

SDTS (Spatial Data Transfer Standard)

SLAR (Sidelooking Airborne Radar)

TM (Thematic Mapper)

URISA (Urban and Regional Information Systems Association)

GIScience

GIS for Everyone, Second Edition
Now everyone can create smart maps for school, work, home, or community action using a personal computer. Revised and expanded second edition now includes ArcExplorer™ 2 and 3 and more than 500 megabytes of geographic data. ISBN 1-879102-91-9

The ESRI Guide to GIS Analysis
An important new book about how to do real analysis with a geographic information system. *The ESRI Guide to GIS Analysis, Volume 1: Geographic Patterns and Relationships* focuses on six of the most common geographic analysis tasks. ISBN 1-879102-06-4

Modeling Our World
With this comprehensive guide and reference to GIS data modeling and to the new geodatabase model introduced with ArcInfo™ 8, you'll learn how to make the right decisions about modeling data, from database design and data capture to spatial analysis and visual presentation. ISBN 1-879102-62-5

Hydrologic and Hydraulic Modeling Support with Geographic Information Systems
This book presents the invited papers in Water Resources at the 1999 ESRI International User Conference. Covering practical issues related to hydrologic and hydraulic water quantity modeling support using GIS, the concepts and techniques apply to any hydrologic and hydraulic model requiring spatial data or spatial visualization. ISBN 1-879102-80-3

Beyond Maps: GIS and Decision Making in Local Government
Beyond Maps shows how local governments are making geographic information systems true management tools. Packed with real-life examples, it explores innovative ways to use GIS to improve local government operations. ISBN 1-879102-79-X

ESRI Map Book: Applications of Geographic Information Systems
A full-color collection of some of the finest maps produced using GIS software. Published annually since 1984, this unique book celebrates the mapping achievements of GIS professionals worldwide. ISBN 1-879102-60-9

The Case Studies Series

ArcView GIS Means Business
Written for business professionals, this book is a behind-the-scenes look at how some of America's most successful companies have used desktop GIS technology. The book is loaded with full-color illustrations and comes with a trial copy of ArcView® GIS software and a GIS tutorial. ISBN 1-879102-51-X

Zeroing In: Geographic Information Systems at Work in the Community
In twelve "tales from the digital map age," this book shows how people use GIS in their daily jobs. An accessible and engaging introduction to GIS for anyone who deals with geographic information. ISBN 1-879102-50-1

Serving Maps on the Internet
Take an insider's look at how today's forward-thinking organizations distribute map-based information via the Internet. Case studies cover a range of applications for ArcView Internet Map Server technology from ESRI. This book should interest anyone who wants to publish geospatial data on the World Wide Web. ISBN 1-879102-52-8

Managing Natural Resources with GIS
Find out how GIS technology helps people design solutions to such pressing challenges as wildfires, urban blight, air and water degradation, species endangerment, disaster mitigation, coastline erosion, and public education. The experiences of public and private organizations provide real-world examples. ISBN 1-879102-53-6

Enterprise GIS for Energy Companies
A volume of case studies showing how electric and gas utilities use geographic information systems to manage their facilities more cost effectively, find new market opportunities, and better serve their customers. ISBN 1-879102-48-X

Transportation GIS
From monitoring rail systems and airplane noise levels, to making bus routes more efficient and improving roads, this book describes how geographic information systems have emerged as the tool of choice for transportation planners. ISBN 1-879102-47-1

More ESRI Press titles are listed on the next page ➤

ESRI® educational products cover topics related to geographic information science, GIS applications, and ESRI technology. You can choose among instructor-led courses, Web-based courses, and self-study workbooks to find education solutions that fit your learning style and pocketbook. Visit www.esri.com/education for more information.

ESRI Press ■ 380 New York Street ■ Redlands, California 92373-8100
www.esri.com/esripress

Other books from ESRI Press

The Case Studies Series CONTINUED

GIS for Landscape Architects
From Karen Hanna, noted landscape architect and GIS pioneer, comes *GIS for Landscape Architects*. Through actual examples, you'll learn how landscape architects, land planners, and designers now rely on GIS to create visual frameworks within which spatial data and information are gathered, interpreted, manipulated, and shared. ISBN 1-879102-64-1

GIS for Health Organizations
Health management is a rapidly developing field, where even slight shifts in policy affect the health care we receive. In this book, you'll see how physicians, public health officials, insurance providers, hospitals, epidemiologists, researchers, and HMO executives use GIS to focus resources to meet the needs of those in their care. ISBN 1-879102-65-X

GIS in Public Policy
This book shows how policy makers and others on the front lines of public service are putting GIS to work—to carry out the will of voters and legislators, and to inform and influence their decisions. *GIS in Public Policy* shows vividly the very real benefits of this new digital tool for anyone with an interest in, or influence over, the ways our institutions shape our lives. ISBN 1-879102-66-8

Integrating GIS and the Global Positioning System
The Global Positioning System is an explosively growing technology. *Integrating GIS and the Global Positioning System* covers the basics of GPS technology and presents several case studies that illustrate some of the ways in which the power of GPS is being harnessed to the depth of GIS: accuracy in measurement and completeness of coverage. ISBN 1-879102-81-1

GIS in Schools
GIS is transforming classrooms—and learning—in elementary, middle, and high schools across North America. *GIS in Schools* documents what happens when students are exposed to GIS. The book gives teachers practical ideas about how to implement GIS in the classroom, and some theory behind the success stories. ISBN 1-879102-85-4

ESRI Software Workbooks

Understanding GIS: The ARC/INFO® Method (UNIX®/Windows NT® version)
A hands-on introduction to geographic information system technology. Designed primarily for beginners, this classic text guides readers through a complete GIS project in ten easy-to-follow lessons. ISBN 1-879102-01-3

Understanding GIS: The ARC/INFO Method (PC version)
ISBN 1-879102-00-5

ARC Macro Language: Developing ARC/INFO Menus and Macros with AML
ARC Macro Language (AML™) software gives you the power to tailor workstation ARC/INFO software's geoprocessing operations to specific applications. This workbook teaches AML in the context of accomplishing practical workstation ARC/INFO tasks, and presents both basic and advanced techniques. ISBN 1-879102-18-8

Getting to Know ArcView GIS
A colorful, nontechnical introduction to GIS technology and ArcView GIS software, this workbook comes with a working ArcView GIS demonstration copy. Follow the book's scenario-based exercises or work through them using the CD and learn how to do your own ArcView GIS project. ISBN 1-879102-46-3

Extending ArcView GIS
This sequel to the award-winning *Getting to Know ArcView GIS* is written for those who understand basic GIS concepts and are ready to extend the analytical power of the core ArcView GIS software. The book consists of short conceptual overviews followed by detailed exercises framed in the context of real problems. ISBN 1-879102-05-6

ESRI Press publishes a growing list of GIS-related books. Ask for these books at your local bookstore or order by calling 1-800-447-9778. *You can also shop online at* www.esri.com/gisstore. *Outside the United States, contact your local ESRI distributor.*

ESRI Press ■ 380 New York Street ■ Redlands, California 92373-8100
www.esri.com/esripress